PORSCHE 356 1952-1965

Compiled by
R.M. Clarke

ISBN 0 946 489 246

Distributed by
Brooklands Book Distribution Ltd.
'Holmerise', Seven Hills Road,
Cobham, Surrey, England

BROOKLANDS ROAD & TRACK SERIES

Road & Track on Austin Healey 1953-1970
Road & Track on Corvette 1953-1967
Road & Track on Corvette 1968-1982
Road & Track on Ferrari 1968-1974
Road & Track on Ferrari 1975-1981
Road & Track on Fiat Sports Car 1968-1981
Road & Track on Jaguar 1974-1982
Road & Track on Lamborghini 1964-1982
Road & Track on Lotus 1972-1981
Road & Track on Maserati 1952-1974
Road & Track on Maserati 1975-1983
Road & Track on Mercedes Sports & GT Cars 1970-1980
Road & Track on Porsche 1972-1975
Road & Track on Porsche 1975-1978
Road & Track on Porsche 1979-1982

BROOKLANDS MUSCLE CARS SERIES

American Motors Muscle Cars 1966-1970
Camaro Muscle Cars 1966-1972
Capri Muscle Cars 1969-1983
Chevrolet Muscle Cars 1966-1971
Dodge Muscle Cars 1967-1970
Mini Muscle Cars 1961-1979
Plymouth Muscle Cars 1966-1971
Muscle Cars Compared 1966-1971
Muscle Cars Compared Book 2 1965-1971

BROOKLANDS MILITARY VEHICLES SERIES

Allied Military Vehicles Collection No. 1
Jeep Collection No. 1

BROOKLANDS BOOKS SERIES

AC Cobra 1962-1969
Alfa Romeo Spider 1966-1981
Armstrong Siddeley Cars 1945-1960
Austin Seven 1922-1982
Austin 10 1932-1939
Austin A30 & A35 1951-1962
Austin Healey 100 1952-1959
Austin Healey 3000 1959-1967
Austin Healey 100 & 3000 Collection No. 1
Austin Healey Sprite 1958-1971
Avanti 1962-1983
BMW Six Cylinder Coupés 1969-1975
BMW 1600 Collection No. 1
BMW 2002 Collection No. 1
Buick Cars 1929-1939
Cadillac in the Sixties No. 1
Camaro 1966-1970
Chrysler Cars 1930-1939
Citroen Traction Avant 1934-1957
Citroen 2CV 1949-1982
Cobra & Replicas 1962-1983
Cortina 1600E & GT 1967-1970
Corvair 1959-1968
Corvette Cars 1955-1964
Daimler Dart & V-8 250 1959-1969
Datsun 240z & 260z 1970-1977
De Tomaso Collection No. 1
Dodge Cars 1924-1938
Ferrari Cars 1946-1956
Ferrari Cars 1957-1962
Ferrari Cars 1962-1966
Ferrari Cars 1966-1969
Ferrari Cars 1969-1973
Ferrari Cars 1973-1977
Ferrari Cars 1977-1981
Ferrari Collection No. 1
Fiat X1/9 1972-1980
Ford GT40 1964-1978
Ford Mustang 1964-1967
Ford Mustang 1967-1973
Ford RS Escort 1968-1980
Hudson & Railton Cars 1936-1940
Jaguar (& S.S) Cars 1931-1937
Jaguar (& S.S) Cars 1937-1947
Jaguar Cars 1948-1951
Jaguar Cars 1951-1953
Jaguar Cars 1955-1957
Jaguar Cars 1957-1961
Jaguar Cars 1961-1964
Jaguar Cars 1964-1968
Jaguar E-Type 1961-1966
Jaguar E-Type 1966-1971
Jaguar E-Type 1971-1975
Jaguar XKE Collection No. 1
Jaguar XJ6 1968-1972
Jaguar XJ6 1973-1980
Jaguar XJ12 1972-1980
Jaguar XJS 1975-1980
Jensen Cars 1946-1967
Jensen Cars 1967-1979
Jensen Interceptor 1966-1976
Jensen-Healey 1972-1976
Lamborghini Cars 1964-1970
Lamborghini Cars 1970-1975
Lamborghini Countach Collection No. 1
Land Rover 1948-1973
Land Rover 1958-1983
Lotus Cortina 1963-1970
Lotus Elan 1962-1973
Lotus Elan Collection No. 1
Lotus Elan Collection No. 2
Lotus Elite 1957-1964
Lotus Elite & Eclat 1975-1981
Lotus Esprit 1974-1981
Lotus Europa 1966-1975
Lotus Europa Collection No. 1
Lotus Seven 1957-1980
Lotus Seven Collection No. 1
Maserati 1965-1970
Maserati 1970-1975
Mazda RX-7 Collection No. 1
Mercedes Benz Cars 1949-1954
Mercedes Benz Cars 1954-1957
Mercedes Benz Cars 1957-1961
Mercedes Benz Competition Cars 1950-1957
MG Cars in the Thirties
MG Cars 1929-1934
MG Cars 1935-1940
MG Cars 1948-1951
MG TC 1945-1949
MG TD 1949-1953
MG TF 1953-1955
MG Cars 1952-1954
MG Cars 1955-1957
MG Cars 1957-1959
MG Cars 1959-1962
MG Midget 1961-1979
MG MGA 1955-1962
MGA Collection No. 1
MG MGB 1962-1970
MG MGB 1970-1980
MG MGB GT 1965-1980
Mini Cooper 1961-1971
Morgan Cars 1936-1960
Morgan Cars 1960-1970
Morgan Cars 1969-1979
Morris Minor 1949-1970
Morris Minor Collection No. 1
Nash Metropolitan 1954-1961
Opel GT 1968-1973
Packard Cars 1920-1942
Pantera 1970-1973
Pantera & Mangusta 1969-1974
Pontiac GTO 1964-1970
Pontiac Firebird 1967-1973
Porsche Cars 1952-1956
Porsche Cars 1960-1964
Porsche Cars 1964-1968
Porsche Cars 1968-1972
Porsche Cars in the Sixties
Porsche Cars 1972-1975
Porsche 356 1952-1965
Porsche 911 Collection No. 1
Porsche 911 Collection No. 2
Porsche 914 1969-1975
Porsche 924 1975-1981
Porsche 928 Collection No. 1
Porsche Turbo Collection No. 1
Reliant Scimitar 1964-1982
Riley Cars 1945-1950
Riley Cars 1950-1955
Rolls Royce Cars 1930-1935
Rolls Royce Cars 1940-1950
Rolls Royce Silver Cloud 1955-1965
Rolls Royce Silver Shadow 1965-1980
Range Rover 1970-1981
Rover 3 & 3.5 Litre 1958-1973
Rover P4 1949-1959
Rover P4 1955-1964
Rover 2000 + 2200 1963-1977
Saab Sonett Collection No. 1
Saab Turbo 1976-1983
Singer Sports Cars 1933-1954
Studebaker Cars 1923-1939
Sunbeam Alpine & Tiger 1959-1967
Triumph Spitfire 1962-1980
Triumph Spitfire Collection No. 1
Triumph Stag 1970-1980
Triumph TR2 & TR3 1952-1960
Triumph TR4 & TR5 1961-1969
Triumph TR6 1969-1976
Triumph TR7 & TR8 1975-1981
Triumph GT6 1966-1974
Triumph Vitesse & Herald 1959-1971
TVR 1960-1980
Volkswagen Cars 1936-1956
VW Beetle 1956-1977
VW Beetle Collection No. 1
VW Karmann Ghia Collection No. 1
VW Scirocco 1974-1981
Volvo 1800 1960-1973
Volvo 120 Series 1956-1970

CONTENTS

ACKNOWLEDGMENTS

The first Brooklands book on Porsche was compiled about twelve years ago and covered much as this one does the early years of the marque. In those days our resources were very limited, in fact our printing plates were made under impossible conditions at home and the outcome was usually not very clear illustrations and from necessity a modest printing run. As a result this book is no longer available.

In recent years we have endeavoured to bring some order into our titles and a higher standard into our printing. At present we cover Porsche in two ways, firstly by model with books on the 911, 914, 924, 928 and all Turbo's and secondly by periods, 1960-1964, 1964-1968, 1968-1972 etc. with no article appearing in more than one book.

As the 356 was the car that established the reputation of the Porsche company after WW2, it was fitting that we should devote a book solely to this important model. The 356 made its debut in June 1948, but by the spring of 1951 only 51 examples had reached customers* and it wasn't until 1952 that the car was being reported on in any detail and why our story begins in that year.

The beautifully preserved car on our front cover is a 1955 model 356 Continental which has had an interesting history and to date has covered some 82,000 miles. It was the show car at the Pan Pacific Auditorium in the fall of 1954 and was also the first 'sun roof' model to be imported into the US.

Its initial owner was Bill Bond the brother of John R. Bond the founder of Road & Track and it has been associated with that famous magazine ever since, as its owner for the past 22 years has been its current Art Director, Bill Motta. He generously not only made the car available, but also posed it in Road & Track's car park, arranged for the sun to shine, took the cover shot with my camera (my only contribution) and at the same time delivered an absorbing lesson in photography. Some days are just better than others. Naturally Bill is another Scorpio!

If you are wondering why you haven't heard of many Porsche Continentals it is because Ford objected to the use of the name and it was subsequently changed to the 'America'.

Over the last twenty years we have compiled over 180 titles such as this for enthusiasts. The outcome has been that about 5,000 stories which might have been lost to today's owners of interesting cars are once again available. We, and in this particular case, Porsche 356 owners, owe a debt of gratitude to the publishers of the leading automotive journals listed below for generously allowing us to include their copyright articles in our reference series – Auto Age, Autocar, Autosport, Foreign Cars Illustrated, Hot Car, Light Car, Motor Sport, Motor Trend, Popular Imported Cars, Road & Track, Speed Age, Sports Cars Illustrated, Sports Car World, Thoroughbred & Classic Cars, Track & Traffic and What Car?

R.M. Clarke

*From Dean Batchelor's excellent book "Illustrated Porsche Buyer's Guide", published by Motorbooks International.

Sam Weill's Porsche 356-4

6

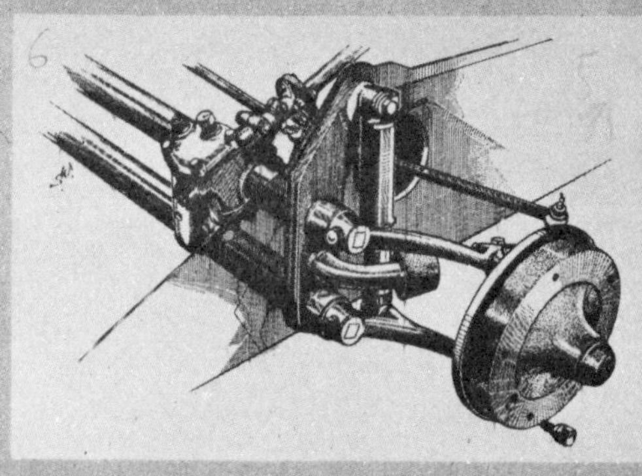

Front suspension . . . trailing arms attach directly to the torsion bars.

Porsche chassis . . . .

Porsche . . .
Technical

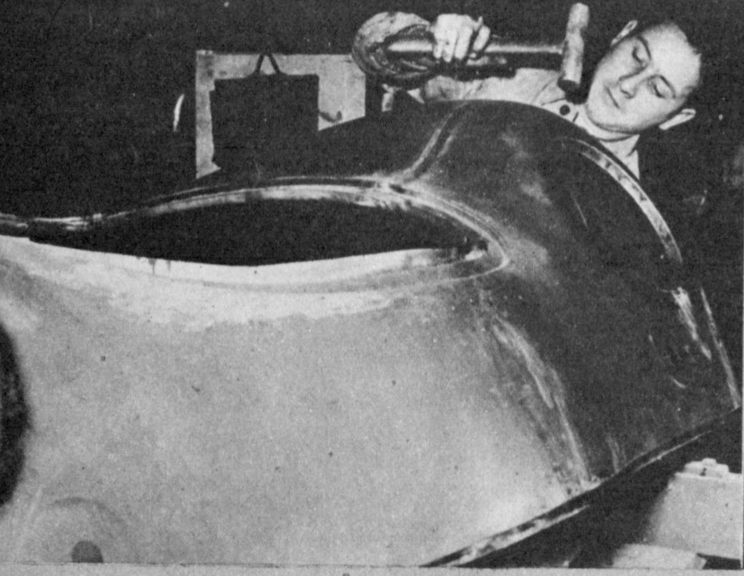

Porsche bodies are individually assembled and finished. If the Porsche were to be hand built in the U.S., price would be triple present $4300.

Each Porsche engine is assembled from the crankshaft up by a single craftsman assigned to that individual powerplant.

Raised hood reveals spare tire, fuel tank, tools, and a limited amount of luggage space. Unsuspecting station attendants are shocked to find engine missing.

Instrument panel features large, well lighted, easy-to-read speedometer and tachometer. Telefunken radio is standard. Clear floor design, made possible by rear engine, gives ample leg room.

Rear of Porsche has such clean lines that only the presence of grille indicates car is rear-engined. Bumpers have rubber rub strip to protect chrome and absorb shock.

Nurburgring . . . . Trenkel at speed in modified Porsche.

Nurburgring . . . . Glochler in his new Glochler-Porsche.

Porsche . . .

Racing

⬆ Dieburg, Germany . . . Six Porsche coupes line up. They dominate 1100 and 1500 cc racing in Germany.

⬇ Montlhery, France . . . one of three record-breaking Porsches which took 17 International records and one World's record. The 1100 cc coupe averaged 100.72 mph for 1000 kilometers (621.4 miles); the 1500 cc coupe set 11 records, among them 3000 miles at 98.95 mph and 72 hours at 94.66 mph; the Glochler-Porsche set three records including 500 kilometers (310.69 miles) at 116.88 mph and six hours at 114.74 mph.

Vero Beach, Florida . . . Max Hoffman at speed in his Glochler-Porsche. He placed first in the 1½ litre class and made second best time of the day.

Porsche trio . . . Lauprecht leads Count Einsiedel thru Dieburg esses.

Le Mans, France . . . competition coupe, which features an all-aluminum body and other modifications, won the 1100 cc class in 1951 and 1952.

8

ROAD and TRACK ROAD TEST No. F-8-52

Porsche Coupe, 356-4

SPECIFICATIONS

Cylinders	4, horiz. opposed	1st gear (overall)	15.93:1
Valves	inclined ohv, pushrod	2nd gear (overall)	9.16:1
Horsepower	65 at 4800 rpm	3rd gear (overall)	5.54:1
Bore and Stroke	80 x 74 mm (3.14 x 2.91 in.)	4th gear " (o'drive)	3.54:1
Displacement	1488 cc (90.7 cu in.)	Curb weight	1850 lbs
Compression ratio	7.2:1	Weight, front	844 lbs
Mph per 1000 rpm	20.7	rear	1006 lbs
Steering lock to lock	2⅛ turns	Weight as tested	2150 lbs
Turning circle	33 ft. 8 in.	Track, front	50¾ in.
Tire size	5.00 x 16	rear	49¼ in.
Transmission	4 speed non-syncromesh; center shift	Overall length	152½ in.
		Overall width	63¼ in.
		Ground Clearance	7 in.
		Seating cap.	2, occasional 3

List price ..... $4208 (East) $4350 (West)

PERFORMANCE

Test conditions—Sea level, calm, dry night, 85° F., Premium fuel

Flying ¼ mile	103.4 mph
Fastest one way	104.6 mph
Standing ¼ mile	18.35 secs.

TAPLEY READINGS

Pulling power	Gear	mph
545 lbs per ton	1st	28
417 lbs per ton	2nd	40
255 lbs per ton	3rd	53
138 lbs per ton	4th	69

Deceleration Rate (Coasting)

32 lbs per ton at	10 mph
44 lbs per ton at	30 mph
78 lbs per ton at	60 mph

ACCELERATION THRU GEARS

0-30 mph	4.8 secs.
0-40 mph	6.7 secs.
0-50 mph	8.7 secs.
0-60 mph	13.8 secs.
0-70 mph	17.5 secs.
0-80 mph	24.4 secs.

SHIFTING POINTS

From	At
1st gear	33 mph
2nd gear	60 mph
3rd gear	85 mph

SPEEDOMETER CORRECTION

Speedometer	Actual
30 mph	27.6
40 mph	36.8
50 mph	46.1
60 mph	55.1
70 mph	67.4
80 mph	75.3
90 mph	85.7
100 mph	97.7

FUEL CONSUMPTION

City driving	27 mpg
Open road	35 mpg
Fuel capacity	13 gal. 1⅓ gal. reserve
Oil capacity	2½ qt.

During the war, while 75% of production for civilian use was stopped and most of the world's economy was geared to Martian needs—we were periodically bombarded by something other than Axis munitions.

That something was the oft repeated prophecy about *The Post War Car.* This fantastic new machine was to be the Dream Car of the Century—completely new in concept. And it was to roll from the production lines as soon as the last battle cry of World War II had died away. It was to be a car such as was envisioned in popular magazines and described by captions which started out "Artist's conception of the car of the future. Note streamlining, etc. . . .".

For various reasons the promised car did not materialize. Instead, in 1946, we were treated to (and happy to get) pre-war design and engineering. In due fairness to the automotive industry, let us state that in the ensuing seven years some advancements have been made. But until the road test of the 356 Porsche coupe, *Road and Track* has found few of the hinted-at engineering concepts actually materialized in production cars.

But, after a turn at the wheel of the new Porsche and a thoro recording of test figures, one is forced to admit that *this* is The Car of Tomorrow. That appears, on its face, to be an exaggerated statement, but experience with the car has given an entirely new driving experience to the test staff. It is safe to say that no car in the history of *Road and Track* has offered so many different and new driving sensations.

In the first place, the car is aerodynamic enough to give the actual feeling that it is (in the words of one of the staff) "slicing thru the air." Secondly, the steering is light enough to allow gradual curves to be taken with only the tip of the thumb resting on the steering wheel. Thirdly, the car is exceptionally economical. If staff member Samuel Weill is to be believed — and we think he is—the Weill Porsche not only covers 27/30 miles on each gallon of fuel in town, but delivers 35 mpg and up on the road — cruising at speeds up to 70 mph. Fourthly, the Porsche is light and streamlined enough to produce virtually unparalleled (for 1500 cc) acceleration figures and a phenomenal top speed.

To recapitulate; the coupe, furnished by John von Neumann of Competition Motors (North Hollywood, California) has an extremely low frontal area for a two (and occasionally three) passenger closed car. Possibly even more important is the smoothness of the airfoil. The profile contour starts at the front of the hood (at bumper height), continues with very little interruption up over the slanting, one-piece curved windshield and over the roof of the car. From there it drops gradually and smoothly to the rear — and the sides of the Porsche are equally well thought out. The result of this design is to give the driver the happy realization that the car needs less "pushing" than others in the 1500 cc class. It goes so *easily.* At 3500 rpm, or 77 mph, the engine seems to loaf along. The driver has no impression of urging the car in order to maintain high speeds. As a matter of fact, with its short stroke, the engine reaches 2500 feet per minute speed somewhere in the neighborhood of a theoretical 130 mph. If we accept the dictum that 2500 fpm is the safe rate of piston travel for maximum cruising speeds in a passenger car, it is difficult to understand how one would go about over-revving the admirable horizontal-opposed four cylinder air-cooled engine.

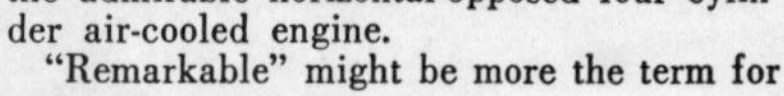

"Remarkable" might be more the term for

Travelers appreciate Porsche's reclining seat. ➧

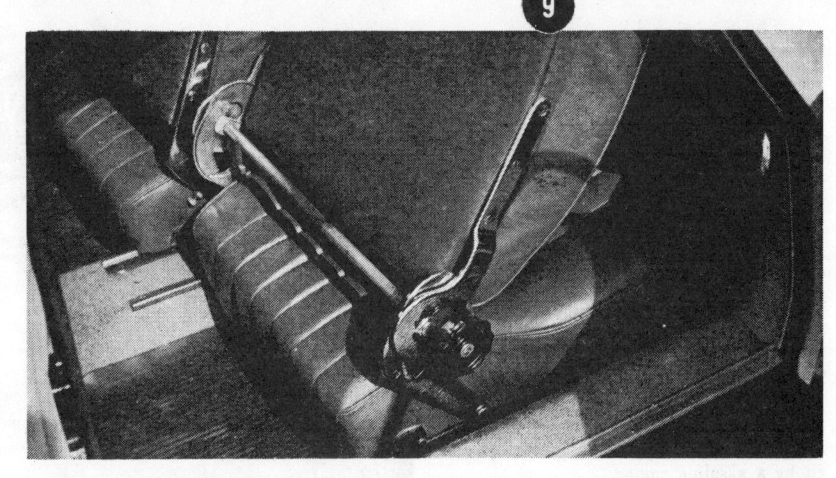

the Porsche engine—with its aluminum-bearing Hirth crankshaft, roller bearing connecting rods, chrome-plated forged aluminum cylinders, and practically all-aluminum construction. The engine, because of this light metal and absence of liquid for cooling, (and 2½ quart oil capacity) is reputed to weigh but 160 lbs.—including all accessories . . . generator, exhaust headers, etc. This lightness permits engine mounting behind the rear axles without disturbing to any great extent the front/rear weight ratio. As far as this balance was concerned, the car manifested none of the "dangerous" handling characteristics so often attributed to rear-engined cars. With the gasoline tank and the 30 pound spare wheel mounted in the luggage compartment under the hood (plus the weight of the driver well forward), the Porsche seems to behave perfectly on the road. It is difficult to diagnose its characteristics as either under or oversteer. If any, the former is present to a happy degree.

To further investigate the handling of the coupe, there is a lightness in the controls of this little 83 inch wheelbase German car which comes as somewhat of a shock on first acquaintance. Anyone accustomed to the steering of "normal" cars is inclined to over-control; much the same tendency as is experienced when first taking the controls of a light plane. And speaking of planes, John von Neumann, who accompanied *Road and Track* on the all-night trek to and from the test strip (and actually drove the car on acceleration and high speed runs), likened the sensations brought on by driving the 356 to those felt in a DC-6 pilot's seat. And it is actually true that the driver feels more as if he were airborne than bound to the highway. It is not poesy to say that the remote thrumming of the happy overhead valve engine in the rear is similar to the outboard engines of the well-known four-engined plane —tho less intense. With its light controls and the dropping off of the gently sloping hood; curved windshield and aircraft type comfort in the bucket seats . . . one may well imagine a plane about to become skyborne. (For those who may feel anxious at the analogy, the curve of the Porsche hood increasingly loads the front end, until at 60 mph there are some 160 pounds of extra air pressure pushing downward on the front wheels. It will definitely not become airborne, or even lose front tire adhesion as speed increases!) To cap the aircraft illusion, the sound of the engine when the throttle is backed off from, say, 70 to 80 mph causes the broad highway in front of you to turn magically into a long landing strip. The feeling is hard to shake off. It is that realistic.

The designers (see Porsche story, page 32) have accomplished many things in the design. In the first place, turbulence between the underside of the car and the road is greatly minimized by the smoothness of what appears to be a "belly-pan," but is in actuality the floor of the car—with protrusions and indentations filled in to give a flat surface from stem to stern. Because not only the engine, but the gearbox is rear mounted, no space is wasted in the "cab" or under the body. All gears and drive units are neatly confined to the rear. This aftward location of the engine and drive train not only serves to cut down the distance over which the power must travel to reach the driving wheels, but also exhibits the tendancy to exhaust engine and gearbox noises backward and away from the driver and passenger—contrary to what happens in the forward-engined, centrally-transmissioned vehicle . . . not to mention the neat dispersal of engine fumes, heat and exhaust gases—all of which are carried off to the rear, without ever menacing the car's occupants.

von Neumann, who so brilliantly won the 1500 cc class at Torrey Pines Road Race in his competition Porsche coupe (with the top cut away), and who proudly welcomed Bob Doidge as *he* came across the finish line in first place with his stock Porsche 356 (in the Production class), says that a worry expressed by most potential Porsche buyers is that the small grille above the engine in the rear "won't let enough air in." Johnny's stock reply is that with the Porsche it isn't a question of letting air in—the huge blower atop the engine takes care of that—but a question of letting the air out. The performance of the car more neatly answered the question, for the little two seater not only failed to show any heating distress, but surprised *Road and Track* when it cooled considerably while idling after each test run.

Dr. Porsche, as you remember, is credited with the development of the laminated torsion bar for use in place of the more conventional coil or leaf spring, so it is not unusual to find that type of suspension on all four corners of the new German import. "Torsion" bars (and the word is advisedly set off in quotes because coil springs also exercise a torsional movement and may in that sense be construed as "torsion bars") . . . torsion bars are usually a solid member of high tensile strength. What the great genius Porsche "invented" was a laminated metal torsion bar which cost less to produce than the conventional bar. The laminated bar will stand greater stresses than a solid bar of the same quality material—hence a greater safety factor is attained at a lesser cost. It is these laminated torsion bars which are found on the Porsche 1500 cc coupe—both front and rear. The front suspension is independent with parallel trailing arms and the rear independent via swing axles. The 356, like the Mercedes 170 sedan (*Road and Track*, February 1952), had an excellent ride over rough roads—tho the car in question was obviously sprung and "shocked" more firmly than the sedan—for a more sports car type ride. But both have the same tendancy to "walk" on a smooth straight highway. It was discovered, with the Mercedes, that any undue increase of tire pressure brought on this walking, while recommended pressures tended to cancel it out. The same may be assumed of the Porsche—in fact was guaranteed by von Neumann, who had the tires pumped up extra hard for the tests. Be that as it may, independent rear suspension is undoubtedly the development of the future. Cost considerations seem to be the only factor today which keeps this device from appearing on all first rate cars.

NOTES AND COMMENTS

The interior finish is excellent. When you see it you will decide for yourself. One glance at the handsome interior and the smart instrument panel will make up your mind. And the passengers seat . . . it is fully reclining . . . adjustable by an easy-to-reach knob . . . marvelous on long trips.

The amount of distortion in the curved windshield was disturbing. This may be a particular batch of faulty glass and may not appear on your Porsche. Three-way switch turns off fuel supply, or cuts in reserve. Firewall—in this this case behind driver—is undersealed, isolating occupants from gear and engine noise. Radio, heater, standard equipment. Crankshaft dynamically and statically balanced. Transmission non-syncromesh but quite simple to operate. Porsche has so many stellar virtues, owners will be glad to put up with some inconvenience of getting in and out of seats. —DEARBORN

PORSCHE STORY

The history of the little 356-4 Porsche (pronounced porsh-eh), the car which created such a profound impression on the *Road and Track* staff, begins back before the turn of the century. It was then that Ferdinand Porsche (born the son of a tinsmith in 1875) used his practical knowledge of electricity to design his first automobile—the Lohner - Porsche - Electrochaise. This car, which had an electric motor in each front wheel, was the hit of the Paris Fair of 1900. A later model incorporated a generator driven by a gasoline engine.

In 1905 Porsche joined the old Austro-Daimler firm, where, among other projects, he designed his first race car, the Prince-Heinrich-Wagen. After brilliant work with this firm on mechanized equipment and aircraft engines during World War I, Porsche again turned to automotive design, producing a series of twin ohc race cars and the still-remembered type A.D.M. sports car of 1927. Turning to Daimler-Benz, Dr. Ferdinand was partially responsible for the famous Mercedes SS and SSK models.

After leaving the firm in 1929 and spending a year with Steyr, Porsche founded (near Stuttgart) an independent designing firm. A team of designers followed the great little man thru development of not only the Auto-Union race cars and torsion bars, but Wanderer and Volkswagen cars.

(left to right) Walter Glockler, race car builder; Max Hoffman, U. S. Porsche distributor; and Ferry Porsche, son of Dr. Porsche.

Porsche was the type of engineer who threw himself heart and soul into each minute facet of design and construction. Porsche realized the importance of aerodynamics as early as 1909; and it was in these early years that he first began to visualize a small economical family sedan for the working man—the dream which was finally realized in the Volkswagen.

It was this study of wind resistance which led Porsche to the development of the highly streamlined "record" Auto-Unions, and the fabulous 6-wheeled Mercedes-Benz intended to attack Cobb's land speed record in 1940.

When Hitler decided every German worker should have an automobile, he dreamed up the "Volkswagen" (translated "people's car" and pronounced folks-va-gen) and put the best resources of Germany on the project (under the leadership of Dr. Porsche).

After 23 months of development, the Volkswagen was ready. Meanwhile, Hitler had used the car as a political lever to extract money from the German public—by taking advance payments on the yet unproduced automobile. But the Volkswagen which appeared in response to the huge sums subscribed was for military use only . . . a jeep.

Professor Ferdinand Porsche gives final instructions to Bernd Rosemeyer just before his record-breaking run in the Auto Union.

(The post-war Volkswagen is Germany's best selling car and threatens to outsell all other small cars in Europe.)

French Occupational Forces took charge of the Porsche plant after the war, and completely dismantled it. Meanwhile, Dr. Porsche languished in the hands of the French authorities—in France.

It is a little known fact that the Porsche car was evolved by Herr Doktor while he lay in custody for two years. Even tho he was ill in body, his active mind went on with plans. During this period, the firm remained faithful to Porsche. Head Engineer Obering, who had been with Porsche since 1913, and Romenda, head of the bodyworks, went on building and designing (in Gmünd, Austria) under the leadership of Porsche's son, Ferry Porsche.

In 1947, *Road and Track's* European Correspondent, Millanta, brought together Porsche agents, Hruschka and Abarth, and Signor Dusio, whose Milano (Italy) factory was engaged in producing the Fiat-based Cisitalia. The result of the meeting was the highly advanced Cisitalia four-wheel-drive Grand Prix car*, which the Austrian firm designed. The Porsche family thus earned the money necessary to pay the French authorities for Porsche's release. The car was actually built and Porsche was released in time to make technical changes in design, but financial problems prevented it from being completed. The Cisitalia is probably the most advanced race car ever constructed, with 450 hp developing from its 12-cylinder, rear - located, horizontal - opposed, 1½ - litre, twin supercharged engine. The lone Cisitalia is still untested, in Argentina.

The actual building of the Porsche—the evolvement of Porsche's prison dream—took place in a defunct aluminum plant in Austria during the winter of 1947-48; and the prototype was so enthusiastically received that more were built and shown in a German auto show. Then the firm moved back to Stuttgart in order to realize greater production. Because the Occupational Forces did not see fit to release the plant they had taken over after the war, Porsche was forced to accept bids for bodies and various parts, Reutter bodyworks being one of the first-class firms employed in this manner.

It is interesting that the Porsche (as tested by *Road and Track*) reaches a top speed of 103 mph with 65 hp. When it is realized that a "conventional" sports car of the Porsche size requires 100 hp or more to attain the same top speed, some idea may be obtained of the 356's aerodynamic efficiency.

As of December 1951, the company was producing but six cars a day and since then little has been done to increase this figure. It is the opinion of Ferry Porsche (who took over management after his father's death in 1951) that vehicles of the Porsche quality cannot be mass-produced.

*See **Road and Track, page 18, March 1951.**

Testing air flow on the Porsche coupe.

For overseas service the smoothly streamlined Porsche coupé has now been given stronger bumpers with over-riders.

NEW CARS DESCRIBED

PORSCHE MODIFICATIONS

SYNCHROMESH GEAR BOX, LARGER BRAKES AND THREE SIZES OF ENGINE

SEVERAL improvements have been incorporated in the latest Porsche, the Type 356, which is now being produced in the Western Zone of Germany. Most important is the introduction of a new gear box with synchromesh for all four speeds. Although Porsche use some Volkswagen parts in engines, steering and suspension, the new gear box is not derived from the synchromesh unit recently introduced for the Volkswagen. It is, in fact, developed from a design which the late Dr. Porsche evolved for the Cisitalia racing cars in 1946-47, and incorporates features which he patented at that time. The synchronization of the gears is effected not by the conventional cones, but by an intermediate servo-synchronizing ring.

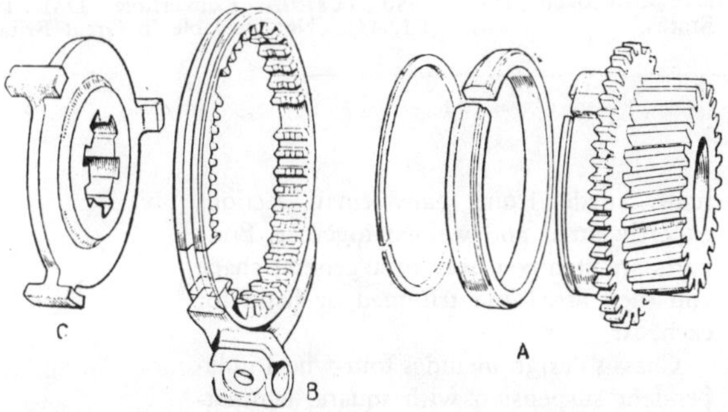

Details of the synchromesh assembly. The gear A carries the servo synchro ring, which is held in place by a circlip. The synchro clutch sleeve B revolves with the spider C, which is splined to the driven shaft. The clutch ring first slides over the servo ring, which adjusts the speed difference.

Each pinion has its own servo ring, which revolves at the same speed as the gear and is held in place by a circlip. The ring itself is rather similar in form to a piston ring. It has a bevelled edge leading to a pressure face approximately 8mm wide and is made of hardened chrome molybdenum steel. It is said to suffer very little wear, as the stresses are distributed uniformly over a large surface. When the clutch ring is moved towards the gear by the gear change fork it first encounters the spring servo ring, which is either braked or accelerated, cancelling out the speed difference between the gear box output shaft to which the clutch unit is splined and the constant-mesh gear, which is revolving in mesh with the

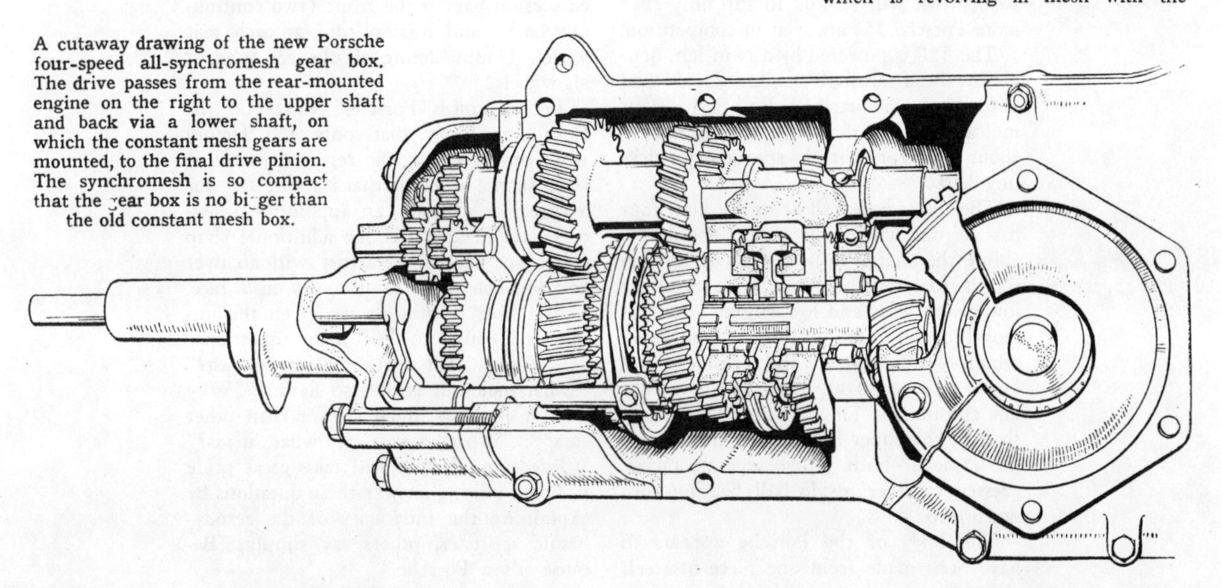

A cutaway drawing of the new Porsche four-speed all-synchromesh gear box. The drive passes from the rear-mounted engine on the right to the upper shaft and back via a lower shaft, on which the constant mesh gears are mounted, to the final drive pinion. The synchromesh is so compact that the gear box is no bigger than the old constant mesh box.

PORSCHE MODIFICATIONS

pinion on the input shaft. With an ordinary synchromesh mechanism, the pauses required are longer at the higher rotational speeds, because the braking or acceleration of the gears takes longer at higher speeds. The spring servo ring, however, creates a servo effect from the moment it is loaded by the synchro-clutch sleeve and speeds up the change so that the interval required for gear engagement is cut down to the minimum regardless of rotational speeds involved.

This ring construction also saves space, and it is said that the new gear box is no longer than the four-speed non-synchronized box which it replaces. At present the new box is available only on new cars and cannot be supplied for fitting to existing models.

Three different sizes of engine are now available in the Porsche. A 1.1-litre engine of 40 b.h.p., a 1.3-litre of 43 b.h.p. and a 1.5-litre which, in standard form, delivers 54 b.h.p. For competitions, there is also a version of the 1.5-litre engine which delivers a maximum of 69 b.h.p. All the engines are flat-four air-cooled units developed from the Volkswagen engine, but they employ many special Porsche features. The crankshafts are new and the cylinder heads are of a special design. The valves are still operated by push-rods from a single camshaft, but by rearrangement of the valve rockers a revised combustion chamber shape approaching the hemispherical has been incorporated.

Perhaps the most interesting innovation is the use of aluminium cylinder blocks. These are not fitted with liners, but the cylinder bore is formed by the deposition of chromium plating directly on to the aluminium, a method which is already in use on some German motor cycles.

With the 1,100 c.c. engine the car is said to have a maximum speed of 87 m.p.h. with an average fuel consumption of 35-40 m.p.g.; the 1.3-litre engine raises the maximum to 90 m.p.h., the fuel consumption rising to about 33-37 m.p.g., while the standard $1\frac{1}{2}$-litre engine gives a maximum speed of some 97 m.p.h. with a consumption around 30-35 m.p.g. The tuned 1,500 engine raises the maximum speed to about 110 m.p.h., as has been repeatedly demonstrated in sports car racing, and with this power unit the fuel consumption claimed is 23-25 m.p.g.

New Brakes

The brakes have been improved to match the higher performance now available. They are Lockheed operated and are two-leading-shoe at the front. Drum diameter has been increased from $9\frac{1}{2}$ to $11\frac{1}{4}$in and the effective lining area is raised from 84 sq in to 124 sq in. The drums are of light metal with cooling fins and have special cast-iron inserts. The new brakes are based on those used on the cars which scored such a decisive victory in this year's Liége-Rome-Liége trial, and are now fitted on all models as standard equipment.

Penetration into overseas markets is having its effect on the Porsche, as it does on other cars, and the clean aerodynamic coachwork has now been afflicted with separate bumpers and over-riders. The standard coachwork styles are a fixed-head coupé and a convertible, but a few open two-seater roadsters have also been delivered to the United States.

SPECIFICATION

Engine.—Flat four, air-cooled, in unit with clutch, gear box and final drive. **1.1-litre,** 73.5 × 64 mm (1,086 c.c.). Compression ratio 7 to 1. 40 b.h.p. at 4,000 r.p.m. Maximum torque 53 lb ft at 3,200 r.p.m. **1.3-litre,** 80 × 64 mm (1,286 c.c.). Compression ratio 6.5 to 1. 43 b.h.p. at 4,000 r.p.m. Maximum torque 62.8 lb ft at 2,600 r.p.m. **1.5-litre,** 80 × 74 mm (1,488 c.c.). Compression ratio 6.5 to 1. 54 b.h.p. at 4,400 r.p.m. Maximum torque 74.3 lb ft at 2,700 r.p.m. **1.5-litre Super.** Compression ratio 8.2 to 1. 69 b.h.p. at 5,000 r.p.m. Maximum torque 79.4 lb ft at 3,600 r.p.m. All engines have push-rod o.h.v., twin downdraught carburettors, aluminium cylinders with chromium-plated bores. Cooling by belt-driven fan.

Transmission.—Dry single-plate clutch, four-speed all-synchromesh gear box. Spiral bevel final drive. Overall ratios 3.54, 4.95, 7.7 and 13.9 to 1. Reverse 15.5 to 1. Axle ratio 4.375 to 1.

Suspension and Steering.—Independent front by double trailing arms and double transverse laminated torsion bars. Independent rear by swing axles connected by single trailing arms to transverse round torsion bars. Telescopic dampers all round. Worm steering gear.

Brakes. — Ate - Lockheed four - wheel hydraulic; two-leading-shoes at front. Mechanical hand brake on rear.

Wheels and Tyres.—5.00-16in tyres on steel disc wheels secured by set bolts.

Dimensions.—Wheelbase 6ft $10\frac{11}{16}$in. Track 4ft $2\frac{13}{16}$in (front), 4ft $1\frac{1}{4}$in (rear). Overall length 12ft $7\frac{9}{16}$in. Width 5ft $5\frac{5}{16}$in. Height 4ft $3\frac{3}{16}$in. Ground clearance $6\frac{3}{8}$in. Weight, dry, 1,640lb.

Price.—Type 356, 1.3-litre coupé DM 9,980 (£875). Convertible DM 11,950 (£1,043). Not available in Great Britain.

Because It's a Porsche

CONTINUED FROM PAGE 33

you have a machine that will place second to few in the 1500 cc class. But what happens when you add 40 horses and subtract 320 pounds from body and chassis? That's a story that will be told in full only after more Porsche 55s are seen in competition.

The 550 is powered by a twin-fan, double-overhead-camshaft engine capable of putting out about 110 hp. Early trials indicate that the 550 will be turning up about 7000 rpm with its speedometer tickling 140.

The large-bore, short-stroke flat four has an unusually slow piston speed, enabling the engine to be revved up at high speeds for hours on end with resulting low piston wear and low engine heat. Removable aluminum cylinders (you don't rebore—just take off the cylinders and install new ones) are plated with vanadium chrome for longer wear. The crankshaft of the Super Porsche is a mechanical masterpiece—built in five separate pieces, pressed together inside ball bearing connecting rod ends.

The body of the Porsche appears to have been made from one piece of steel; closer examination reveals that it is actually made from many small sections expertly fitted and welded together. Body parts are stamped only to a general shape and then fitted and trimmed by hand for each car.

Chassis design includes four-wheel independent suspension with square, laminated torsion bars at the front (two continuous bars) and torsion rods at each rear wheel. Double-acting shocks are used at all wheels.

Because most Porsche owners on the Continent insist that only the factory work on their cars, the repair department is on such a schedule that it's necessary for a customer to make an appointment several days in advance. An additional shop is set up where the factory will go over any Porsche prior to its entry in a race.

The Porsche has always given the uninitiated cause to ask many questions: "How does that little car go so fast?" "What makes it corner so flatly?" "Why does it look so much sleeker than other cars?" "What makes it what it is?" While some people still take great pride in being able to answer these questions by explaining the intricacies of the aerodynamic sportster, others say simply, "Because it's a Porsche."

—T. C. Countryman

Liege to Rome and Back

Porsche triumphed in one of Europe's most famous rallies, run over 3230 miles where drivers face the added handicap of everyday traffic.

The winning Porsche in the Alps during the exhausting 3230 mile long distance run.

Rugged mountain driving took the greatest toll of entrants. Only 24 cars were able to finish out of 106 starters.

Text and Photos

By GUNTHER F. MOLTER

The International Liége-Rome-Liége Rally is the famous long distance race on open roads in Europe. Since the roads are not blockaded, contestants face the added handicap of coping with everyday traffic.

The most recent rally was one of the most difficult. In less than 90 hours the drivers had to cover 3230 miles across Belgium, France, Italy and Germany. The fact that of 106 cars, which started from Liége, only 24 reached the finish line tells the story.

The route crossed the Alps and included 33 mountain passes, 16 at more than 6000 feet.

The winning car was a 1500 CC (91 cubic inch) Porsche 356 SL, lent to the drivers—two Germans, Helmut Polensky and Walter Schluter—by Miss Gilberte Thirion, a Belgian race driver and holder of the flying kilometer (0.6214 mile) record for a Porsche in Europe. In addition to the efforts of victors, the entire Porsche team completed the race. ☆ ☆

THE PHENOMENAL PORSCHE

by JOHN GOTT

WHILST inspecting the cars competing in the 1949 Coppa d'Oro Internazionale della Dolomiti, I was greatly intrigued by a beautifully streamlined fixed-head coupé of a new type. The coachwork, despite the bruises and bumps almost inseparable from a "works" experimental hack, was of superb aerodynamic design and the whole car gave an immediate impression of lightness and efficiency. The driver was loath to discuss technical details, but I gathered that the car was based upon certain Volkswagen parts, and was the brain-child of the late Dr. Ferdinand Porsche, who was not only the designer of the "P"-Wagen, alias Auto Union, but also largely responsible for the Volkswagen.

Although the car, not being *au point*, performed indifferently in the race, I was convinced then—and am still more convinced now!—that a lightweight coupé, planned by a racing designer of genius around basic components proved and tested by literally millions of miles in civilian and military use, must turn out to be an outstanding combination. I did not, however, anticipate that this car would be developed to the point of being able, in Rally trim, two up, to do 112.5 m.p.h. over the flying kilometre and cover the same distance from a standing start in 35.1 seconds, in 1½-litre form.

After a successful début in smaller national competitions, the Porsche began to show its potentialities in International events in 1951. At Le Mans a 1,086 c.c. car easily won its class at an average speed of 73.6 m.p.h. (this being faster than the average returned by the 1,500 c.c. class winner), turning in a lap of 87 m.p.h. In the toughest Rally in the calendar, Liége-Rome-Liége, a Porsche was classified equal third and again easily won its class. In the Tour de France, a Porsche took fourth place and another class win. For the Freiburg hill-climb at the end of the season, the new 1,488 c.c. engine was fetched out, and the 1,500 c.c. class taken with a time only bettered by two cars in the unlimited sports car class and by the redoubtable Stirling Moss in the Formula 3 class. To round off a rather shattering season, International Records were taken at Montlhéry at speeds varying from 115.69 m.p.h. for 1,000 kilometres to 94.66 m.p.h. for three days.

Through the initiative of Charles Meisl, now with Colborne Garage, Ltd., Porsches appeared at the 1951 Earls Court Show, thereby taking another record as being the first German car to appear there since the Hitler War. Although, as a sellers' market was then ruling, the cars were locked and jealously guarded, it was possible to gather a little technical information, as well as to admire the excellent finish.

At that time there were two catalogue models, with 1,086 c.c. and 1,286 c.c. engines, but the 1,488 c.c. engine was expected to be catalogued in the spring of 1952. The engine was a rear-mounted air-cooled flat four, giving off the quite modest output of 40 and 44 b.h.p., according to the c.c., at 4,000 r.p.m. The weight of the car, a two-seater coupé in fixed and drop-head forms, was quoted at about 15 cwt. This, combined with efficient aerodynamic shape, enabled the cars to pull gear ratios of 3.54, 5.54, 9.17 and 15.95 to 1 on 5.00 x 16 tyres. Front suspension was by independent trailing arms and torsion bars, rear suspension by swing axles, trailing arms and torsion bars. The petrol tank, holding about 10½ gallons, and the spare wheel, reposed under the bonnet where the engine is housed in a more orthodox car.

As a matter of interest, one of the Show cars is alleged to be in the experimental shop of a well-known British manufacturer, whilst Harry Sutcliffe owns another and the third is used as a means of transport between home and office by Gordon Claridge, who used to operate a 328 BMW before the Hitler War.

Although success in one year does not by any means guarantee a similar success in the following year, 1952 was to prove even more successful for the "marque" than 1951. Indeed, by the end of the year it was beginning to be taken for granted that Porsches would win their class in every International Rally or Sports Car Race in which they were entered, and it was a matter for comment if they did not either win or finish high in the General Classification.

Owing to difficulties about coachwork dimensions, they did not enter in force for the Rallye Monte-Carlo. However, in Rallye del Sestriere they took second and third places in General Classification and easily won their class. In the Rallye Soleil, they sank to fifth place

in general classification, but completely swamped the 1,500 c.c. class, in which they filled the first six places.

In the Mille Miglia Porsches took the first three places in the 1,500 c.c. Gran Turismo Class, and first place in the 1,100 c.c. Series sports class. This did not go unnoticed in a country famous for small, fast sports cars. In the Coppa Inter-Europa prior to the Italian G.P., Count Johnny Lurani took a Porsche into first place in the 1,500 c.c. class at an average of over 86 m.p.h.

The Tulip Rally, in which Ray Brookes and I drove my 1947 H.R.G., gave me a chance to meet the Porsches in direct competition for the first time. Nine Porsches started in the 1,500 c.c. sports class, opposed by a mixed bag of M.G.s, a Gran Sport Simca, an old Riley Imp and the H.R.G. On the run from Brussels to Rheims, whilst cruising at around 75 m.p.h., I was startled to be passed by a Porsche which had left Brussels 28 minutes behind me. However, I was not therefore unduly shocked when Porsches took eight of the 10 class placings on the timed climb on the Ballon d'Alsace, and Von der Muhle's time was only beaten by eight cars in the whole entry, all of these being in the unlimited sports class. The "outsiders" were Van der Lof, the eventual class winner, on his cut-down, and very stark M.G. TD, and myself. Even Ted Lund's very fast M.G. TD, ex-Thornley "works" car of Silverstone and Dundrod fame, could not live with the Porsches. I counted myself extremely lucky to finish fourth at Zandvoort, behind Van der Lof and the Porsches of Von der Muhle and Nathan, and certainly would not have done this had not three Porsches displayed a defect in road-holding and crashed into the sand-dunes.

Observing these unpleasing sights from the immediate rear, I came to the conclusion that when driven into corners at the extremely high speeds of which the cars were capable, the combination of swing axle, rear mounted engine and low general weight caused a breakaway at racing speeds to come without warning and some violence. The more experienced drivers could control this, but it certainly prevented the less skilful from exploiting the full potentialities of their cars.

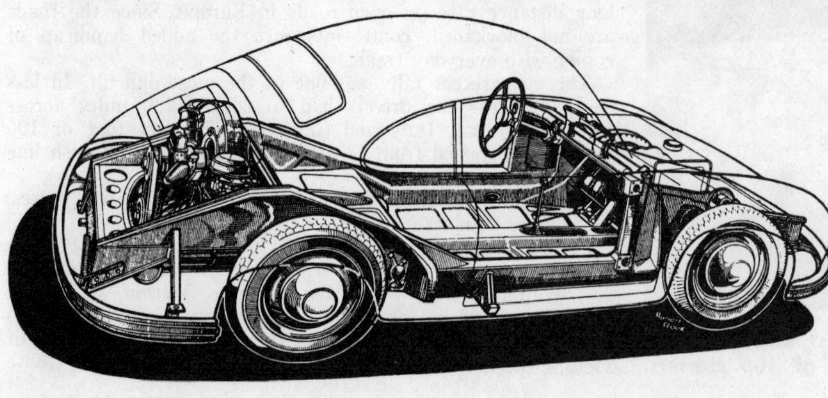

VOLKSWAGEN-BASED: Cutaway drawing (from L'Automobile*) of the Porsche coupé, with rear-mounted, flat-four, air-cooled engine, and all-round independent springing.*

FAMILIAR VIEW: Von der Muhle's Porsche at Zandvoort during last year's Tulip Rally.

However, Porsches now started to sweep the Rally board. They annexed first, third and fourth places in the Lisbon Rally, first second and fourth places in the Rally of the Midnight Sun, and fourth and fifth places in Rallye-Travemünde. In all these events, the cars also took the class and team prizes.

At Le Mans, the 1,500 c.c. Porsche was disqualified, when leading the class, for making a pit stop with the engine running, but not before it had lapped at over 91 m.p.h. The 1,086 c.c. car, driven by Veuillet and Mouche, finished 11th, at 76.47 m.p.h., easily winning the class at a speed above the record for the 1,500 c.c. class.

As a result of my experiences in the Tulip, I knew that no M.G. TD in anything like standard form could begin to cope with the Porsches, and whilst I felt that an H.R.G. was a better match, I did not think it fair to call upon my old car, which had already finished four Alpines, for another supreme effort in this most strenuous of events.

Accordingly, I felt myself very lucky to be able to drive Jupiter HAK 366, which had won the 1½-litre class in the 1951 T.T. and performed nobly at Le Mans and Silverstone, in the 1952 Alpine. I was even more fortunate to have as a team-mate Tommy Wise on the prototype Jupiter, which had won the class in the 1950 Le Mans.

The Alpine was, however, only a qualified success for the Porsches. Although finishing first and second in the 750-1,500 c.c. class, both cars were penalized, and in this class alone was no coveted Coupe des Alpes awarded. As my H.R.G. had made an unpenalized run in 1951, the set average speed was not impossible, and as the Porsches climbed very much faster, it was obvious that they were not rapid enough down-Alp. In fact they suffered from brake troubles, and could not live with the Jupiter on the descents. Thus it was a bitter disappointment when the Jupiter's brake-drums, although quite adequate for the lesser strain of Le Mans and Dundrod, began to break up under the Alpine hammering. This meant that the car had to be withdrawn on the third day, although running perfectly, for the addition of strengthening plates. At that time only the two Jupiters and one Porsche were left unpenalized in the class.

In Liége-Rome-Liége, probably the toughest rally ever run, I was co-driver to Lt.-Colonel H. C. O'Hara Moore in his Le Mans Frazer-Nash. For many miles we ran in company with the "works" Porsches. Up to about 90 m.p.h. their performance was little inferior to that of the 'Nash. The brakes, a new type now in production as a result of their success in this strenuous event, did not suffer from fade, and modifications to the suspension allowed the "works" drivers at any rate adequately to control any incipient breakaway. Polensky and Schluter, the winners, were never out of the first three in the 12 timed sections, which was a fantastic performance considering that they were competing on level terms with some of the world's finest sports cars, many of treble their cubic capacity, driven by acknowledged Alpine experts. Porsches swept the board, scoring first, third and fourth places, both team prizes and the almost inevitable class win.

This amazing result was, however, the high watermark of the cars' 1952 performances. Although taking first and second places in the 1,100 c.c. class in the Tour de France, the cars were unplaced in General Classification.

In conversation after Liége-Rome-Liége, I gathered some interesting facts about the cars. Although models straight off the production line are often entered, as in the Rally of the Midnight Sun, four light alloy bodied cars form the spearhead of the "works" attack. These, however, only weigh some 80 lbs. less than the production models and have covered many thousands of competition miles. The crankcase is now about all that remains of the original Volkswagen engine. The cylinders are light alloy, with chromed bores. The crankshaft is a fully counterbalanced Hirth unit. Pushrods operate valves, of which the inlet is of greater diameter than the exhaust, in a perfect hemispherical head. Compression ratio is 8.5/1, and in this form the cars give off around 75 b.h.p. Cooling is assisted by a large turbo-fan, and the absence of a radiator does much to bring down the weight. The Volkswagen gearbox was unable to cope with the increased power, and a new and most ingenious type of synchromesh gearbox has been recently designed and fitted. A gearbox of this type may be incorporated in the new Formula 2 Maserati, which is a good indication of its worth.

The "works" drivers stressed that not only was competition found to be the best test bed, but that the successes gained in competitions all over the world were invaluable publicity. Of set design, the cars were entered in any event in which they were thought to stand a good chance, and, apart from the victories previously mentioned, class wins were scored in the Agadir Road Race, Eifelrennen, 12 Hours Race at Hyères, Coppa Dolomiti and Bridghampton. In American events, indeed, drivers who used to appear at the wheel of Abingdon products have now gone over to Porsches.

Porsches have only twice appeared, prior to Harry Sutcliffe acquiring his car in the late autumn of 1952, in English events. Neither Merck nor Buschmann, on 1,286 c.c. cars, performed outstandingly in the R.A.C. Rally, in which the manœuvring tests did not suit their gearboxes. Bannister's showing in a race at Castle Combe was also indifferent. English drivers should not, however, be misled by these performances and should, indeed, be glad that Porsches are only available in England to diplomats and Americans, neither of whom are likely to be interested in competition.

Tolworth and Idle are, however, well aware of this foreign supremacy in a class which used to be regarded as an English preserve, and some interesting developments are expected from these factories in 1953.

TECHNICAL SPECIFICATION OF THE PORSCHE, TYPE 356

ENGINE (All models) : Rear mounted, horizontally opposed " flat four," air-cooled. Twin Solex carburetters. Hemispherical head.

	1.1 litre	1.3 litre	1.5 litre	1,500 Competition
Bore	73·5 mm.	80 mm.	80 mm.	80 mm.
Stroke	64 mm.	64 mm.	74 mm.	74 mm.
Capacity	1,086 c.c.	1,286 c.c.	1,488 c.c.	1,488 c.c.
Compression Ratio	6·5/1	6·5/1	7/1	8·5/1
Max. B.H.P. developed	40	44	55	70/75
at R.P.M.	4,200	4,400	4,400	5,000

Clutch : Single dry plate. **Axle Ratio :** 4.375/1. **Gear Ratios :** *First*, 3.18/1 ; *second*, 1.76/1 ; *third*, 1.13/1 ; *top*, 0.815/1 ; *reverse*, 3.56/1. **Tyre size :** 5.00×16. **Brakes :** Hydraulic. Twin master cylinder on front, single master cylinder on rear wheels. Mechanically operated handbrake on rear wheels. **Turning Circle :** 34 feet. **Wheel base :** 6 ft. 11 in. **Track :** Front, 4 ft. 2¾ in. ; rear, 4 ft. 1¼ in. **Dry Weight :** 1,640 lb., approx. 14¾ cwt. **Overall length :** 12 ft. 7½ in. **Overall Height :** 4 ft. 3¼ in.

	1.1 litre	1.3 litre	1.5 litre	1,500 Competition
Catalogue Maximum Speed	87 m.p.h.	90 m.p.h.	96 m.p.h.	109 m.p.h.
Catalogue Petrol Consumption	35 m.p.g.	33 m.p.g.	31 m.p.g.	24 m.p.g.

Catalogue Acceleration for 1.5 litre car: s.s. 0–50 m.p.h., 9·1 seconds; s.s. 0–100 k.p.h. (62 m.p.h.), 14·6 seconds.

TESTING THE

Text and Photos

By M. B. CARROLL, JR.

SPEED AGE FEATURE WRITER

Speed Age's sport car authority gets his hands on an extremely small automobile and learns he can get 30 miles per gallon, 111 MPH and corner at suicidal speed.

Airflow over the sloping front adds up to 150 pounds pressure at speed and increases the stability of the Porsche.

MOST sport car owners follow a rather set pattern of which I suspect I'm about to become a victim. The first car is almost invariably a MG. In a year or so the MG is tearfully traded on a Jaguar XK-120 (the eternal search for more 'dig') and then, when the bankroll recovers, the Jag is traded for a Porsche.

The steps are logical enough. The MG offers an inexpensive and wholly practical means of investigating the advantages and disadvantages of a thoroughly fascinating hobby; the Jag offers glamour and performance that will outshine anything you are apt to meet on the road; the Porsche offers. . . . . well, let's see what it offers.

Looks are a controversial point whether the subject is cars or women. Ferry Porsche, son of Dr. Ferdinand Porsche, has said that the body form is dictated only by aerodynamic principles. With the Mercedes-Benz 300SL body style dictated by the same precepts, it is not surprising that the body forms are similar. Wide and extremely low, almost completely devoid of chrome, this car's functional beauty grows on me. Conversely, my mother, a woman of strong opinions, refused to ride in the Porsche because of its ugliness.

Most impressive and least subject to opinion are the car s amazing capabilities. Its comfort and riding qualities on any road would amaze even a Bentley owner. Its handling, with its 83-inch wheelbase, seem to make the car responsive to the driver's thoughts rather than to any conscious effort. A steering ratio that gives 2⅛ turns lock-to-lock, combined with the car's light weight and apparently friction-free steering mechanism, make for a responsiveness to the wheel that is dangerous until you become thoroughly accustomed to it. Once you get the knack of it, the Porsche corners with no apparent effort at suicidal speeds. The trick seems to be to turn the wheel a wee bit in the direction of the curve and then change your mind, straighten out the wheel, apply power and sail through in a perfect 4-wheel drift. Phil Walters gave a beautiful lesson in how to handle the Porsche when he won the final event at the Brynfan Tyddyn Road Races near Wilkes-Barre, Pa., last year driving Briggs Cunningham's roadster.

I was fortunate enough to be in charge of one of the safety stations spotted at intervals around the course. Mine was a particularly good spot because it commanded a view of a fast bend as well as a slow 90-degree corner. The sight of that low roadster, remaining perfectly level as Phil drifted it through the fast bend at an impossible speed, was something I will never forget. John Gordon Bennett, who placed second in his blown MG, also gave a grand lesson in how to handle the MG but it was obvious the Porsche was the better-handling car. And, with all this cornering ability, the car rides as softly as anyone could desire thus proving that comfort and stability can be combined in one automobile.

Part of the secret lies in the suspension. The front wheels are connected to laminated torsion bars through trailing links and the rear wheels are independently sprung through swing axles and another laminated torsion bar. Perhaps the 4-wheel independent suspension contributes to the feeling of oversteer in hard corners. This is a characteristic of every car I've driven with 4-wheel independent suspension; now that I'm familiar with it I like it, but I insist this is not an automobile to be raced until the driver has had time to familiarize himself with the handling characteristics.

The functional aerodynamic design means that the car moves through the air with a total lack of wind noise and an occasional glance at the speedometer is necessary if one is to keep from creeping over the legal speed limits. This lack of wind noise was illustrated to me when I took the cornering pictures. Although the car was traveling well over 70 MPH when it passed me, there was no sound except a faint murmur from the exhaust. Comparatively, my Ford sounded very loud when it came around the same bend at a slightly slower speed. I had hoped to get a picture of the Ford following the Porsche through a fast bend but the acceleration of even the relatively low (55 HP.) powered model we were using for roadability and handling tests was too much for my beloved flivver. The car is equipped with Fordamatic and I later berated Jerry Kiser, who had been driving it, for not starting in Lo. "Hell," said Jerry, " I did start in Lo." Next car around the bend contained the local gendarmes—'nuff said. No picture.

The car I used for most of my tests was a blue convertible owned by Ned Freeman of Westfield, N. J. Ned, a manufacturing engineer with Minneapolis-Honeywell in Philadelphia, drives to work every day—160 miles round trip—and he never spends more than three hours on the road. This, mind you, without using the N. J. Turnpike. Ned says he gets over 30 miles per gallon and, as my own tests showed an average of 32.1, I see no reason to doubt him. This is excellent mileage, particularly when one considers that the car has over 8,000 miles on the odometer and that it has never had more than routine maintenance.

Jersey weather being what it is in the winter I had ample opportunity to check both the heating and defrosting equipment and the car's ability to keep out rain. The system

The gas tank and spare wheel under the hood confuse service station attendants. Luggage is stored behind the seat.

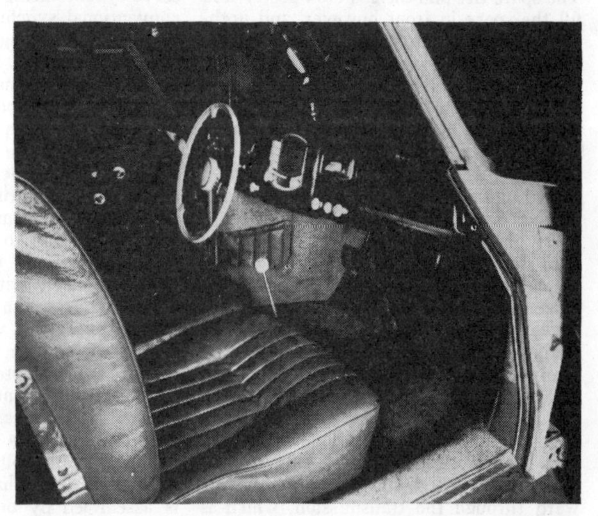

The interior is comfortable and roomy. The radio and heater are standard equipment in all convertible models.

Owner Ned Freeman drifts his Porsche through a bend at 70 MPH. On such high speed turns the car rides perfectly level. The low weight of the Porsche, suspension and the light, quick steering make for terrific responsiveness to the wheel.

worthy of some explanation. Leading forward from each side of the rear-mounted air-cooled engine are two ducts that parallel the chassis side members. Cool, fresh air is drawn in through the grille in the rear deck, circulated around the engine by the cooling fan, and forced forward through the ducts to the passenger compartment and the windshield defroster outlets. Simple and very effective. I like it.

Interior appointments are excellent; the fully adjustable leather seats leave nothing to be desired and the leg room is ample for anyone except a circus giant. On the right side of the passenger's seat is a huge black knob which changes the angle of the seat back from vertical to near horizontal, just the thing for a sleepy passenger.

Luggage space is about on a par with that of the Jaguar with any difference favoring the Porsche. Most of the luggage room is behind the seats and there's space for a few small packages under the hood. The spare tire and the gas tank pretty well fill the space in front however, and nothing is as much fun as to pull into a service station and ask to have the tank filled and the oil checked. When the attendant gives up looking for the gas cap you can tell him to forget it and check the oil. Up comes the hood and the mystery of the missing gas cap has been solved but the question of 'where's the engine?' is next.

By placing the air-cooled engine at the rear, Dr. Porsche gave himself a chance to indulge in some really top-flight engineering. Design-wise, I'm convinced this is the most advanced thing on wheels. The engine is a flat opposed 4-cylinder job, assembled as a unit with the transmission and differential. Removal of the engine and complete drive train takes less than one hour and the engine alone is so light that even I can heft it with only reasonable amount of effort. The drive is forward through the transmission, which is mounted ahead of the differential, then back through the transmission to the differential and the swing axles. This gives a minimum of unsprung weight and is responsible for the handling and riding excellence.

The engine has been turned out in so many sizes and stages of tune that I can't keep track of them all. For sure, it is available in the 1,100 cubic centimeter (67.1 cubic inch) size with an output of 44 HP and in two stages of tune in the 1,500 cubic centimeter (91.5 cubic inch) size producing some 55 HP in the standard jobs and 70 HP in the Super models. The competition model roadster comes with an overhead cam engine that turns out some 84 HP and will, as I discovered the hard way, run the ears off a stock Jaguar.

The original models delivered in this country were equipped with a straight-cut, non-synchromesh gearbox and displayed their capabilities only when handled by experts. The new models are equipped with a new synchromesh box which uses servo-synchronizing rings instead of the conventional synchronizing cones.

To get the thing down to readily understood terms, it shifts faster than the ordinary synchromesh box and, furthermore, all four speeds are so equipped that you can even shift down into first without a struggle if the course or the road conditions demand it. Surprisingly enough, this feature alone makes the newer models from one to two seconds faster from zero to 60 MPH.

The Super and the competition roadster are fitted with roller-bearing connecting rods which reduce friction some 40%. Couple this with the fact that the engine has to turn more than 5000 RPM before the accepted cruising speed of 2,500 feet per minute of piston speed is reached and it's easy to see why nobody has yet blown up a Porsche engine. Heads would probably roll in the factory if such a thing were to happen as each engine is assembled by one man and he is responsible for its performance and reliability.

Performance is far from sluggish. I ran my performance tests on the Super model owned by George Tilp of Short Hills, N. J. and, after finding that I couldn't beat the synchromesh, turned in some amazing figures for a 1,500 cubic centimeter automibile. Zero to 60 ran 10 seconds flat, the standing quarter was turned in 17.6 seconds, and top speed was a fat 111 MPH.

For my money, the best bet of the Porsche line is the America coupe. This is the standard 1,500 cubic centimeter model with the new synchromesh transmission, new and larger brakes (11½ inch diameter instead of 9½ inches) and an engine that turns out 55 HP. With an overall weight of 1,840 pounds this car will provide all the performance anyone could reasonably ask for and will deliver over 30 miles to the gallon in the bargain. Price is $3,395 delivered in New York and it's a bargain.

Owning one of these cars in my home state of New Jersey is a troublesome and expensive problem because of the attitude of the personnel connected with the compulsory inspection set-up.

The Porsche is fitted in the windows with a new type of safety glass that disintegrates upon impact. Instead of flying into a million sharp pieces, however, this glass breaks into smooth-edged pieces that do not cut. I believe the Army Air Force is about to adopt this glass because it is safer and less likely to injure personnel than is the conventional type of safety glass. The sovereign state of New Jersey will not pass this glass and demands its replacement with the more dangerous conventional type.

All in all, the car offers more roadability, handling ease, comfort, economy of operation, and performance than anything I've yet driven in its price or displacement class. I don't think you could ask for anything else.

SPECIFICATIONS

Engine:

4-cylinder, horizontal, opposed, air-cooled. Displacement 1,488 cubic centimeters (90.7 cubic inches). Bore 3.14 inches. Stroke 2.91 inches. Overhead, pushrod operated valves. Compression ratio 7.2/1. 70 HP at 4800 RPM.

Transmission:

4-speed synchromesh, center shift. Ratios (overall): 1st—15.93/1; 2nd—9.16/1; 3rd—5.54/1; 4th—3.54/1.

Dimensions:

Wheelbase—83 inches. Front track—50.75 inches. Rear track—49.25 inches.

Overall height—51 inches. Overall width—63.25 inches. Overall length—152.5 inches.

Weight—1,860 pounds. Distribution front/rear—46/54.

Steering:

2⅞ turns lock-to-lock.

Performance—Super model.

0—60 MPH (average of three runs)—10.0 seconds.

Standing ¼ mile (one run)—17.6 seconds.

Maximum speed (average of two runs)—111.2 MPH.

Price:

America—55 HP—$3,395

Super—70 HP—$4,284

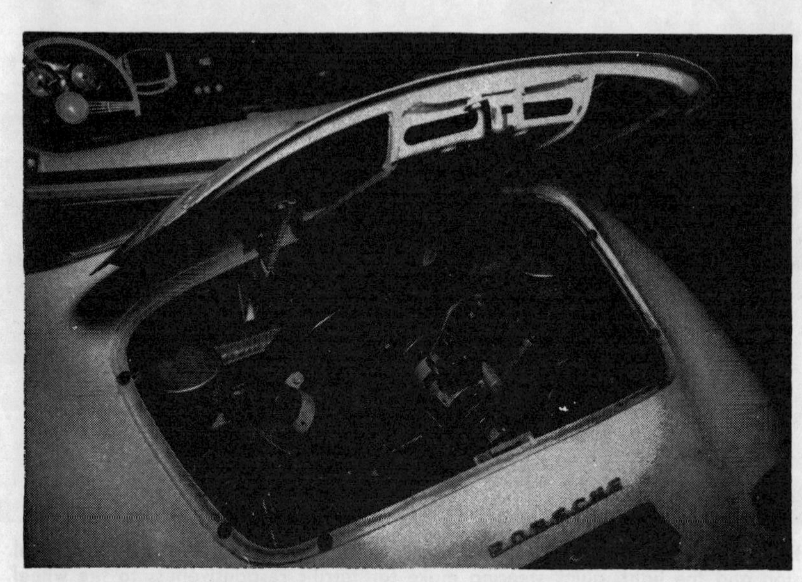

The 4-cylinder, horizontally-opposed engine of the Porsche is air-cooled.

ROAD IMPRESSIONS OF A STRIKING CONTINENTAL 1½-LITRE SPORTS COUPÉ

THE TYPE 356

PORSCHE

IN the recent Alpine Rally, Porsche cars were placed first, second and fourth in general classification and occupied the first eight places in the 1,600-c.c. class. Such sweeping success is clearly no mere fortuitous accident and it was therefore with particular interest that we recently accepted the invitation of Colborne Garage, Ltd., of Ripley, Surrey, to carry out an all-too-brief test of a Type 356.

The photographs on this page indicate something of the compact and graceful lines of the 1½-litre coupé. What, perhaps, is not made so clear are the proportions of the car: it is only 4 ft. 3 in. high, has a ground clearance of a little over 6 in. and an overall length of 12 ft. 7½ in. The kerb weight is approximately 15 cwt. and the standard 1,500-c.c. engine develops 55 b.h.p. at 4,400 r.p.m. Thus, with 37 b.h.p. per litre and a power/weight ratio of 30 lb. per horse-power, high performance can confidently be anticipated.

Nor is that confidence misplaced. The car has remarkable accelerative powers (0 to 50 m.p.h. through the gears in 9.1 sec.) and a maximum speed of about 95 m.p.h. It is very

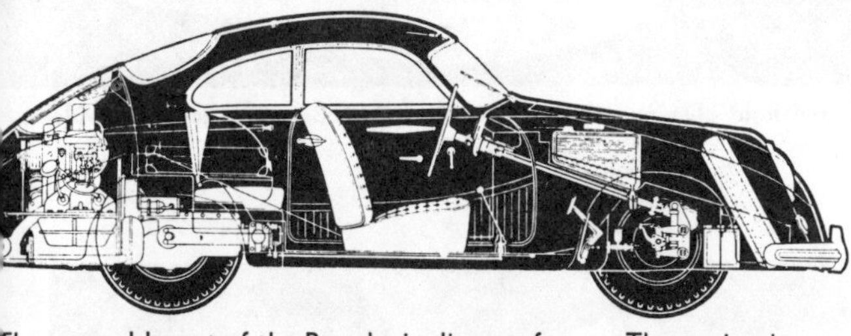

The unusual layout of the Porsche in diagram form. The engine is mounted well behind the rear wheel centres, whilst only petrol tank and spare wheel occupy the "bonnet."

definitely a sportsman's car and its unusual design demands a handling technique quite different from that employed with a car in which the greater proportion of the weight is concentrated in the forward half of the chassis. The Porsche engine (air-cooled, horizontally opposed o.h.v. four-cylinder) is mounted behind the rear axle, a position which creates marked oversteer characteristics. This is by no means a fault of design but it is something which the new owner must learn before the combination of wet road and over-enthusiastic right foot induces the Dreaded Spin.

If the steering has a " different " feel at cruising speeds of 70 m.p.h. it is completely accurate, high-geared, light and free from wander. The comfort of one's ride over all types of surface is one of the delights of the car. Pot-holed roads, adverse cambers, hump-backed bridges, can all be taken with confidence at a fast cruising speed.

The Porsche chassis is of steel pressings welded to form a rigid box structure. Front independent suspension is by two square, laminated torsion bars and telescopic dampers, and there is a single torsion bar at the rear. The individual seats are deep and shaped to give firm support when cornering and

The Alpine Rally-winning Porsche of Polensky and Schlueter on the Col de l'Iseran. Five cars of this type finished the course without loss of marks.

there is ample head and leg room for the tallest crew.

A feature of the car which makes an immediate appeal to the enthusiast is the four-speed gearbox which is a joy to use. Changes both up and down can be made rapidly and there is little gear noise: indeed, the rear-mounted engine is far less obtrusive than one might expect.

Forward visibility could hardly be improved: both front wings are in full view but there is some slight distortion of vision through the curved screen at its outer edge. Incidentally, the standard gear ratios have been chosen with unusual skill. Second can produce nearly 40 m.p.h. and third a gait in the middle seventies—and with the minimum of delay; furthermore, fuel consumption is better than 30 m.p.g.

The Porsche is one of those cars the smoothness of which belies its speed. There is no buffeting of wind around the screen pillars, no suggestion of a hard-working engine behind the rear seats, and the driver has no half-suppressed anxiety that the road is a trifle narrow at 80 m.p.h. The brakes are light in action, powerful if need be, and free from fade, and the 5.00 by 16 tyres have to be sorely tried before they howl.

Imported motorcars today are expensive and the price of the Porsche (inclusive of duty) is £1,971. A costly vehicle, certainly, but one which offers performance exceeded in this country only by the sports racing car. Perhaps this renewed foreign competition in the 1½-litre field might result in increased interest in a class which the British, not so long ago, regarded almost as their own.

Driver's side: instruments are neatly arranged, forward vision is unobstructed and the seats give full support on fast corners.

ROAD
AUTO AGE
TEST

Even with bumpers removed, top raised and light aluminum racing seats, Roadster's beautiful lines are evident.

"One-man" top which folds neatly behind seats, can be raised in 45 secs. if tackled by two pairs of hands.

Performance Data

Acceleration through gears
- 0-30 mph: 4.1 secs.
- 0-50 mph: 7.4 secs.
- 0-60 mph: 9.3 secs.

Third gear 30-50 mph: 4.6 secs.

Standing 1/4 mile 17.9 secs.

Maximum speed 110 mph

Brake test
- From 30 mph: 24 ft. 6 in.
- 45 mph: 50 ft. 5 1/2 in.
- 60 mph: 89 ft. 3 in.

Fuel consumption
Under all traffic conditions and including all tests: 26.4 mpg.

Speedometer correction
- At 40 mph, read 40 mph: No error
- 50 mph, read 48 mph: 4% slow
- 60 mph, read 59 mph: 1.6% slow

Easy winner of Class F in 100-Mile Cumberland main event, writer's Porsche gets rain-soaked checkered flag. Car ran faultlessly.

THE PORSCHE ROADSTER

By JOHN BENTLEY

When Phil Walters won the difficult 1952 Brynfan Tyddyn (Pa.) road race in a trim Porsche Roadster belonging to Briggs Cunningham, little did I dream that I would become the eventual owner of this rather unique machine. Walters, on that July weekend, undoubtedly turned in the finest performance of his brilliant amateur career, squeezing out of the nimble Porsche a speed that seemed to. defy all laws of centrifugal force and gravitation. He even contrived a series of power drifts through turns that fully taxed the most experienced drivers of machines with conventionally located engines, and repeatedly used an approach which for most other drivers would have been downright suicidal with a rear-engined car. And that he put more into winning this race than he had ever done before, was obvious from the perspiration that streamed down his face when he pulled in after getting the checkered flag.

The Porsche car, as such, has several times been reviewed and more than once been road tested. It is therefore not my purpose to recapitulate here the peculiarities of construction found in this superb little car which is, of course, the realization of a dream long nurtured by that automotive genius—the late Dr. Ferdinand Porsche. I want to talk specifically about the Porsche 1500 *Roadster,* of which only about a dozen examples are at large in the world today.

After its Brynfan Tyddyn victory, the ex-Cunningham Roadster had a pretty easy life until it came into my hands. It made an appearance at Thompson Track (Conn.) driven by Briggs Cunningham Jr., and again briefly, at the 1952 Watkin's Glen Grand Prix, when the throttle linkage worked loose after two laps. This, I gathered was the sum-total of its racing activities prior to its being offered for sale early in the summer.

Let me say here and now (*Continued on page* 22)

Porsche	GOOD	AVERAGE	SLIGHT	POOR	BLIND SPOT	EXCESSIVE	NIL
Hard cornering—lean							✓
Panic stops—nosedive							✓
Spring rebound—rough road			✓				
Steering wheel road shock			✓				
Steering wheel lost motion							✓
Right front fender vision	✓						
Left windshield pillar vision					✓		
Rear window vision	✓						
Rear view mirror				✓			
Legibility—instruments	✓						
Driver's seating position	✓						
Rear leg room							
Rear head-room							
Restfulness—long trips	✓						
Windshield reflection			✓				
Sun glare protection							✓
Headlight effectiveness	✓						
Wind noise		✓					
Luggage space		✓					
Detail finish	✓						
Running economy	✓						

The Porsche Roadster

that the views I want to express about the Roadster have probably more of an academic than a practical interest for the reader, since (even if you should passionately desire an example of this model and are prepared to shell out $4,600 for it) your chances of getting a new one are practically nil. Only four Roadsters reached this country during the past year and all were quickly sold. Once in a blue moon, and with a sly elusiveness highly frustrating to the enthusiast, one of these gems shows up all too briefly on the used market.

Wherein lies the almost hypnotic attraction that this model holds for the connoisseur of thoroughbred sports machines? If you happen to own either a Series "356" Porsche Coupé, or a 1500 Super Coupé, there is little need for me to explain. The Roadster is the same kind of machine—*only much more so.* It is a lot faster on pick-up; at least 10 mph fleeter at the top end; and—because its 12-in. brakes have to retard about 300 pounds less weight than on the Coupé, they are more fantastic yet.

If you aren't even a sports car en-enthusiast and think that the man who is willing to part with $4,600 for a diminutive job with a hand-hammered aluminum body, a canvas top, and *not even a radio* should have his head examined, then read further.

THE Porsche Roadster will climb up the back of a Modified Jaguar XK120 up to around 70 mph, and will easily outdrag a Cadillac from zero to any speed. Correctly driven, it will outcorner and outbrake absolutely anything in the genuine sports car line, anywhere near its price. And for sheer comfort expressed in terms of long journeys at very high cruising speeds and of really smooth riding over rough surfaces, it is altogether a fabulous small car. Fast progress on cobbled or rippled surfaces, over ruts and bumps, is marked by a total absence of jounce, rebound, pitching, rolling or any of the other tiring manifestations common to conventional suspensions found in much larger and heavier cars. And that's a good deal to say, even for torsion springs.

Before acquiring the Roadster, I had owned two Porsche Coupés, so that I well knew the capabilities of the marque, but this one—if ever any small car did—deserves to be called a midget with ten-league boots. It was the first simile that hit me after I took the wheel; I have had reason to reaffirm that impression.

The model under review, being a 1952 job, does not have the close ratio sychromesh gearbox common to the latest Porsches, nor the sodium valves of the newest series. Second and third gear are a shade too widely spaced for best use of engine power, and the box can hardly be called a silent one in first and second gears. But with even an elementary knowledge of double-clutching, it performs like the proverbial knife through butter. The challenge to correct co-ordination is a highly stimulating and agreeable one. Clutch action is light but positive, and the inclusion of a ball-type throwout bearing enables the clutch to withstand much abuse from the careless or uninitiated.

On my car, the featherlight steering was altogether too free, even for a Porsche, and as a result the car had a tendency to wander at high speeds. This was cured by tightening the two-way adjustment on the steering box, which provides both for end play and the pressure of the nut on the worm gear.

AGAIN, on the last few Roadsters to reach this country (and on all the latest Supers and Americas) the Alfin brakedrums are one-piece jobs with only enough steel backing to anchor the wheel studs. The cast-iron hubs used on the 1952 Roadsters have been done away with, resulting in a weight-saving of about two pounds per drum on each of the front wheels and one pound on each of the rear wheels. Otherwise, braking action is identical and truly amazing. The stopping figures for the car tested give a good idea of the Porsche's braking efficiency, but only by sampling this feature can it really be appreciated. You are doing, say, 80 or 90 mph and you want to stop in a hurry. Just stab the brake pedal two or three times and you find yourself loafing at 25 mph in an incredibly short distance. Seemingly fade-proof, even under the hardest racing use, the Porsche's brakes enable you to go far deeper into the turns than any of the opposition.

In the matter of handling, the Roadster is neither more nor less sensitive than the other models. The reduced weight offers no added handicap to its excellent roadability, *provided* you observe the same basic precautions as with any other Porsche. That is, always get your braking and shifting over and done with before you enter a corner, and go through under moderate power, gently wishing (rather than steering) the car around, to compensate for the marked feeling of oversteer. Follow this simple rule and you can out-corner anything but an OSCA; ignore it and you may find yourself in serious trouble. Roughly handled, the Porsche will swap ends faster than your quickest reactions can anticipate this maneuver.

Tire pressures have a greater effect on the Porsche's behavior than ever is the case with a front-engined car, and for this reason it pays to check your tires frequently. Keep the Roadster's front tires at 21 pounds and the rear ones at 30 pounds (racing) or 26 pounds (touring) and you will have no trouble.

The simple, husky, compact and superbly designed Porsche engine is not noted for quietness of operation. In fact, in its healthiest state of tune this engine can produce the weirdest and most alarming collection of sounds ever to come out of an internal combustion power unit. But once warmed up and in its stride, the rhythmic beat of this engine is a reassuring obligato with a quality so satisfying that it becomes indispensable. Perhaps this stems in part from the fact that the absurdly low piston speed makes it impossible to overwork the mechanism, and if you know anything about Porsches you are con-

PORSCHE ROADSTER SPECIFICATIONS

Number of cylinders	4
Bore	3.15 in.
Stroke	2.91 in.
Displacement	90.76 cu. in. (1,488 cc)
Compression ratio	8.2:1
Maximum output	75 bhp at 5,500 rpm
Bore/stroke	.92:1
Bhp per cu. in.	1.21
Valves	OHVPR
Carburetors	Two Solex Downdraft 40 PBI-C
Transmission	Four speed (Constant mesh third and fourth)
Overall ratios	4th: 3.54:1 2nd: 9.17:1 3rd: 6.53:1 1st: 15.94
Rear axle ratio	4.43
Weight (car tested)	1,581 pounds
Power/weight ratio	21 pounds/bhp
Wheelbase	83 in.
Turning circle	34 ft.
Turns (lock to lock)	2¾
Tire size	5.00 x 16
Tread	(front) 50¾ in; (rear) 49¼ in.
Overall height	51¼ in.
width	65¼ in.
length	152½ in.
Ground clearance (min.)	6 in.

stantly and agreeably aware of it. For example, at 4,000 rpm, piston speed is a mere 1,940 feet per minute—yet 4,000 rpm represents not far off 90 mph in high gear! An MG, on the other hand, has a piston speed of 2362 fpm at the same rpm. With a correctly balanced engine, high rpm's are not even fractionally as destructive as high piston speeds which enormously increase bearing loads and inertia stresses on the rods.

THE Roadster engine (because bulkhead insulation is less important than weight-saving) obtrudes far more than is the case with the Coupe; but to each his own. If you want a true sports car of small displacement with real racing performance, you must be prepared to make some compromise. And if ever a machine deserved the title of true sports car in every sense of the word, it is this Porsche Roadster. You can drive it two or three hundred miles to a race with the top up and in perfect comfort, and once there you can strip off much excess weight in minutes with one wrench.

The windshield (21 lbs. 4 oz.) fastened by two wing-nuts and a center bolt, for which a special box wrench is provided, comes off in about three minutes. The folding canvas top (27 lbs, 8 oz) is sprung in place by two pins and can be lifted out complete in 15 secs. The bumpers (19 lbs 8 oz front; 14 lbs 9½ oz. rear) are held in by four bolts and can be slid out in five minutes, easily. Two small wing-nuts locate the luggage compartment tray, removal of which takes less than 30 secs. That's another 9 lbs 9 oz saved.

A curved plastic racing windshield (which is part of the car's equipment and weighs only 1 lb. 1 oz) can be installed in a few seconds by hand-tightening a couple of small nuts, and the car is ready for racing. It now scales 91 lbs 7½ oz less than when you arrived, but still more can be done by way of weight reduction with even less trouble. Whip out the jack (6 lbs 15 oz); the tool-roll (5 lbs); the hubcaps (4 lbs); the tonneau cover (3 lbs 15 oz), and the right floorboard and rubber mat (4 lbs 12 oz) and the Roadster is unburdened by 115 lbs 17½ oz. Lively as the machine is with its full complement of accessories, the shedding of this much weight makes quite a difference.

You will observe that the car is equipped with two aircraft-type aluminum bucket seats, and not the handsome leather-upholstered armchair seats originally provided. Here again, weight-saving was the reason. The original seats weigh about 38 lbs. apiece (including slide rails), whereas the racing seats tip the scales at roughly 20 lbs each, with cushions. That means a further 36 lbs pruned off, bringing the reduction in ballast to nearly 152 lbs—or the equivalent of an adult passenger. For short courses where maximum acceleration is vital, a featherlight, plastic-upholstered aluminum garden seat is substituted on the passenger side, which fully conforms with FIA requirements. Since this weighs only 3 lbs, the grand total saved is almost 169 lbs—which is the same thing as if I didn't ride in the car but handled it by radio!

The weight indicated in the specifications table includes everything on board and about six gallons of gas, so that in fully stripped condition the Roadster tips only 1412 pounds—giving a power/weight ratio of 18.8 pounds/bhp—which is a shade better than the XK120, but with superior roadholding and braking, and a much smaller frontal area.

The factory stated engine output of 75 bhp is a conservative one as are all Porsche claims. This figure was obtained on the Clayton chassis dynamometer and therefore represents hp delivered *at the road wheels*. Stripped of auxiliaries and run on a test bench, this engine undoubtedly would register over 90 bhp. The advertised bhp of big domestic engines, by the way, is never that delivered at the road wheels, but on the bench. Thus, deflation of the so-called 160 hp engine, for example, on the chassis dynamometer to a mere 135 hp, comes as sad and bewildering news to many proud owners of Detroit Iron.

THE "physiological" make-up of the Porsche Roadster engine, by the way (if one can use such a term), is amazingly similar to that of a human being. In some respects it has inexhaustible strength and stamina, while in others it is as sensitive and vulnerable to abuse as the human body. For instance, you can wind it up repeatedly to 6,000 rpm through the gears without hurting anything; but go just *one* rpm beyond this limit, and it is a mathematical certainty that the hot exhaust valves will hit the piston head—and sooner or later will bend. In the car under review this drawback was overcome by carefully grinding and polishing the entire rocker gear assembly—a tedious and time-consuming job; and by turning the cam followers down on a lathe to conform with Glöckler specifications. The aggregate weight-saving of a few ounces in these vital reciprocating parts, retards valve float by possibly 200 rpm; so that if you do accidentally hit 6,050 rpm, the valves come to no harm. There is no measurable difference in liveliness, but the resultant safeguard is well worth the trouble—if you race.

The other peculiarity of this engine is its acute susceptibility to temperature changes, via the carburetion system. You may be going happily and in splendid tune, when suddenly you enter climatic conditions entailing a rise or drop of 10 deg. Fahrenheit in air temperature. That's all you need. The carburetion is thrown out of whack and the engine quickly loses that fine edge which normally characterizes its performance. If you are touring, it doesn't matter much, except that you can *sense* something is not quite right. But if, on arriving at a road race or hillclimb you ignore these symptoms, they may easily cost you your chances of doing anything.

Here, a procedure which we will term "jet jugglery" becomes imperative. The Roadster comes from the factory with 135 m/m jets; 60 m/m accelerator pump jets and 160 m/m air correction jets. These are a pretty good compromise for normal running, but if the air gets hot (and therefore expands) *less* of it reaches the engine and the mixture is affected. The harder you rev. up, the richer the mixture—which means loss of performance at the top end of the rpm range. Here, the cure seems to be to fit a larger air correction jet—say 170 m/m. Conversely, if you encounter a temperature drop, the air contracts and *more* of it is sucked into the engine. That means a leaner mixture at high rpm, and again loss of power. Go back to the 160 m/m air correction jet. But there is also the main jet to consider. The larger this is (within reason) the cooler the engine runs and the more "dig" it has when you mash down the throttle. It would seem, therefore, that a larger main jet (140-m/m) also is indicated in hot weather, although this (as explained above) is not necessarily the cure. A main jet increase may also do the trick in cold weather, and there is, of course, the humidity factor to consider. Save under extreme conditions one way or the other, it fortunately is not necessary to fool around with the accelerator pump jet, since this only operates up to about 1,500 rpm; but the main and air correction jets can lead you a merry chase while you make calibration runs up the road to determine the optimum settings.

For this reason, when racing I carry a board handy which holds a bewildering selection of jets of every possible dimension. The only other adjuncts required are a wrench (metric), a screwdriver and a monumental amount of patience. But when you do find the right combination for that particular event, the engine booms happily and acceleration almost snaps your head off.

As to plugs, I have found KLG F80's very suitable for all-round use, though no doubt the equivalent heat range in other reliable makes would serve equally well. F80's can even be used for racing, if the event is not too arduous, though I usually switch to colder F100's. However, these plugs are *not* suitable for traffic driving and will foul up unless you rev. the engine far beyond requirements under these conditions. For everyday use, a .024-in. gap gives best results with F80's.

Jets or no jets, the Porsche Roadster is a peerless gem among thoroughbred sports cars. On city streets it has the docility of a family car, yet it has beaten disguised Grand Prix machines costing twice as much.

A typical example of the versatility of this power-packed mechanical marvel was the recent Cumberland (Md.) 100-Mile SCCA airport race. Carrying my wife, myself and a heap of baggage and spares, the Porsche breezed 400 miles to the event, won its class in both the Ladies and Main Event with only a change of plugs, then sped home as sweetly and effortlessly as when we started. At cruising speeds of 60-70 mph, by the way, gas consumption worked out at 29.7 mpg!

Once become a Porsche addict and you have had it as far as any other breed of small automotive machinery is concerned. It is much more than fine design; it is engineering artistry and inspired craftsmanship.

CALLING

This exclusive article by the leading U.S. authority on Porsches tells you exactly how to tune and maintain the World's finest small car and keep it at peak efficiency.

TABLE A—ENGINE CHARACTERISTICS BY SERIAL NUMBERS

Engine Nr	Type	Bore	Stroke	Displ.	Comp. Ratio	BHP at RPM
P10001–up	369	73.5mm	64mm	1086 ccs.	7:1	40–4000
P20001–up	506	80mm	64mm	1286 ccs.	6.5:1	44–4200
P30001–30066	502	80mm	74mm	1488 ccs.	7:1	55–4500
P30067–30750	527	80mm	74mm	1488 ccs.	7:1	60–5000
P30751–up	546	80mm	74mm	1488 ccs.	7:1	55–4400
P40033–up	528	80mm	74mm	1488 ccs.	8.2:1	70–5000

TABLE C—TUNE-UP SPECIFICATIONS

Engine Type	Tappet Clear. inlet	Tappet Clear. exhaust	Ignition Timing	Bosch	Spark Plugs KLG	Champion
369	.008	.006	11/32 in.	W 225 T 1	F–70	L 10 S
506, 502, 546	.004	.004	11/32 in.	W 225 T 1	F–70	L 10 S
527	.006	.004	13/32 in.	W 225 T 1	F–70	L 10 S
528	.006	.004	13/32 in.	W 240 T 1	F–80	L 11 S

TABLE B—SOLEX CARBURETORS AND SETTINGS

Engine Type	Carburetor Type	Venturi mm	Main Jet	Idling Jet	Pump Jet	Correction Jet
369	PBI–32	23	0110	60	50	230
506	PBI–32	24	0115	60	55	240
502, 546	PBI–32	24	0120	55	55	260
527	PBI–32	26	0115	55	55	180
527(altern.)	PBIC–40	26	0115	55	60	160
528(to Ser. # 40127)	PBIC–40	26(29)	0115 (130)	55	60	160
528 (from Ser. 40128)	PBIC–40	26(29)	0107.5 (0117.5)	50	85(110)	160

Note: Figures in brackets represent *basic* settings for competition or racing.

TABLE D—VALVE TIMING AT .015 INCH TAPPET CLEARANCE

	Type 527 and 528	All others	Timing
Inlet opens	42 1/2 deg.	17 deg. 10 min.	Before TDC
Inlet closes	77 1/2 deg.	52 deg. 10 min.	After BDC
Exhaust opens	77 1/2 deg.	52 deg. 10 min.	Before BDC
Exhaust closes	42 1/2 deg.	17 deg. 10 min.	After TDC

ALL PORSCHES

By KARL GRASSOW

The name Porsche has long been associated with new ideas in automobile design. When, early in 1952, the Porsche factory introduced the Type 527 engine it established another precedent. Although previous Porsche models had been outstanding in their respective classes, here was the first standard production car in automotive history that was capable of exceeding a genuine 100 mph with an engine of less than 1,500 ccs. displacement.

The old truth that you cannot get anything for nothing applies also to highly developed sports cars, and the price for retaining the performance that is built into these vehicles includes proper maintenance.

Each Porsche car is supplied with a comprehensive instruction manual at the time of purchase. As is almost always the case, however, certain modifications are made in production during the course of a year, and with the engineering and sales forces forever a step ahead of the printers, literature easily becomes obsolete. Bulletins are then rushed to the dealers but much too often this new information does not reach the individual owners of the cars. A heavy burden of correspondence is thus placed upon the technical personnel of the manufacturer or his representatives and when I was asked to write this article I jumped at the opportunity with the hope that it would provide a medium through which to answer many questions at once.

BHP/RPM AND TORQUE CHART

BHP: 10, 15, 20, 25, 30, 35, 40, 45, 50, 55, 60, 65, 70

RPM X 100: 15, 20, 25, 30, 35, 40, 45, 50, 55, 60

——— 1500 ENGINE TYPE 528 (SUPER)

– – – – 1500 ENGINE TYPE 546 (AMERICA)

The standard Porsche sports car, in Coupé or Convertible form, is designated by its makers as the Model 356. Since the inception of exports to the U.S. in 1950 only minor detail alterations have been made to the chassis and bodies; major modifications being confined to the engines and transmissions. To date, six different engines have been fitted to these export cars and they can readily be recognized by the serial numbers which are located on the crankcase casting directly under the generator. The engine characteristics relating to these various numbers are shown in Table A.

In Porsches, the quoted bhp figures must not be confused with advertised claims of other makes which often have the benefit of manipulations during, and recalculations after, tests in accordance with standards that are laid down by automotive industry groups. Porsche horses have big, hairy legs and the listed figures represent sustained outputs of production engines "as installed;" i.e. equipped with standard accessories, ignition, carburetion, exhaust and cooling systems. It is not infrequent that a Porsche will show 95 to 100 per cent of these values at the rear wheels, when placed on a Clayton Chassis Dynamometer.

Now as to tuning procedure: if the history of a particular car is not known it is always advisable to check the carburetor settings. Various Porsche engines have at different times been fitted at the factory with the Solex carburetors and settings indicated in Table B.

It is important to note that the fuel level in PBIC-40 carburetors is critical when these instruments are installed in Porsche engines. It pays to check this by filling the carburetors with the standard fuel pump, using the engine starter, removing the top bodies of the carburetors and measuring the distance from the upper edge of the float chamber to the fuel level. This should be between

SEQUENCE FOR TIGHTENING PORSCHE CYLINDER HEAD NUTS (SET TORQUE WRENCH AT 22 FT/LBS.)

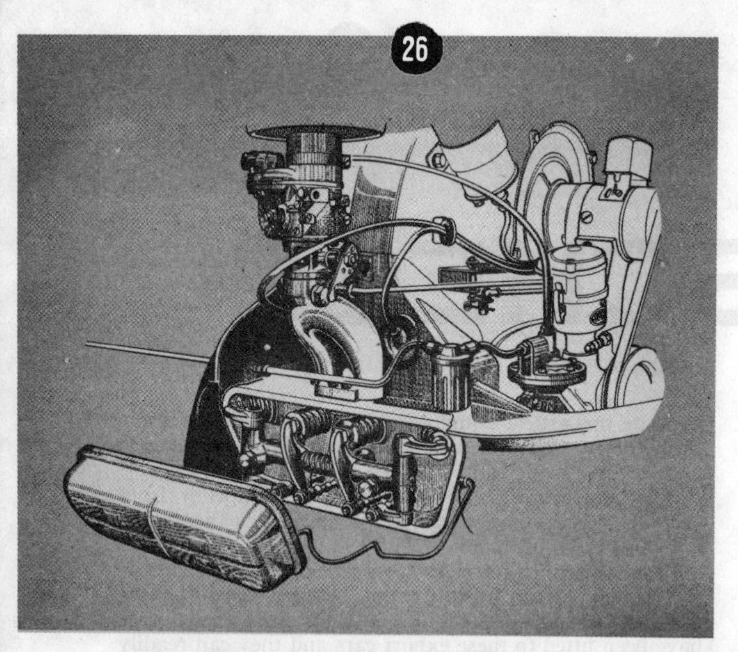

Newer 1500 cc engine types 528 and 546 have tappet adjustment on pushrod side of rockers instead of valve-spring side as formerly.

¾ and 49/64-inch. Adjustments can be made by adding fibre washers under the float needle valve if the level is too high and subtracting if too low.

In tuning a Porsche engine a methodical step by step procedure must be followed (see Table C Tune-Up specifications) in order to obtain satisfactory results and avoid duplications of effort.

Starting to work on the *cold* engine, the tappets must be adjusted first. The adjustment must be made with the piston at the end of its compression stroke (Top Dead Center) in the respective cylinder. A TDC mark for the No. 1 cylinder is provided on the crankshaft pulley. After bringing this mark in line with the split of the aluminum crankcase halves and making certain that the distributor rotor is in the four o'clock position, the tappets of No. 1 cylinder can be adjusted. Following the firing order 1-4-3-2 and turning the engine 180 degrees for each successive cylinder, the job can be finished rapidly. The cylinder positions are:

No. 1—right front; No. 2—right rear; No. 3—left front; No. 4—left rear, facing back of car.

Clean spark plugs of the correct heat range, gapped to .027-in. should now be installed. The distributor contact point gap must be set to .016-in. (46 degrees cam dwell) and the timing checked and adjusted if necessary. Table C also gives the distance in inches on the crankshaft pulley rim where the distributor points should break, before the TDC mark has lined up with the split in the crankcases. To make matters easier it is worthwhile to place a timing mark on the pulley rim, the prescribed distance to the *right* of the TDC mark.

The engine should now be started and brought up to operating temperature, at a steady 2000 rpm in order to avoid sooting of the spark plugs due to possible maladjustment of the carburetors. Under normal touring conditions the oil temperature gauge on the dash will indicate between 70 and 90 Centigrade (approx. 160 to 195 degrees Fahrenheit), and no carburetor adjustments should be attempted unless at least the lower of these figures is shown on the gauge. This is, in fact, the primary function of this instrument. The maximum permissible oil temperature under sustained high speeds or in racing is 120 Centigrade (appr. 250 degrees Fahrenheit). It should be borne in mind, however, that the gauge shows the temperature of the oil after it has returned to the sump, or at its hottest. The oil is picked up by the oil pump at the lowest and coolest point of the sump and forced through the oil cooler before it reaches the bearing, pistons etc. of the engine. Since the oil cooler lies in the direct air blast of the cooling fan a drop in temperature of the oil from 50 to 70 degrees Fahrenheit (depending on ambient temperatures) can be expected there. I thought it well to mention this since several drivers have dropped out of races unnecessarily when the gauge showed temperatures in excess of the prescribed maximum. Actually they still had a useful temperature margin in hand, although this was not indicated by the gauge.

After the engine has reached operating temperature it should be stopped. The carburetor linkage will have expanded somewhat due to the surrounding heat and must be attended to first. Proceed as follows:

1) Back off both throttle stop screws until the throttles are closed.
2) Loosen both nuts on the right side of the long rod that connects the lever of the two carburetors.
3) Check that both throttles can be worked freely by hand.
4) Close throttles completely and, while holding them closed, re-connect the long rod.

The air cleaners should now be removed from the carburetors to check the action of the accelerator pumps. This must be simultaneous. If one is lagging behind, the pump passages and check valves must be cleaned of dirt or obstructions before attempting any adjustment of the pump linkage. It must be under-

TABLE D

PORSCHE TORQUE WRENCH SPECIFICATIONS

ENGINE		
Crank case nuts 10 mm	22 ft/lbs.	
Crank case nuts 8 mm	14 "	
Cylinder head nuts	22 "	
Rocker arm nuts	36 "	
Flywheel bolt	290 "	
Connecting rod bolt 369, 506	36 "	
546	25 "	
Fan nut	72 "	
STEERING		
Steering wheel nut	40 ft/lbs.	
Pitman arm bolt	50 "	
BRAKES & WHEELS		
Brake backing plate bolts	40 ft/lbs.	
Wheel studs (230 mm Drum)	95 "	
Wheel nut (280 mm Drum)	80 "	
TRANSMISSION & REAR AXLE		
Bolts & nuts for case	14 ft/lbs.	
Ring gear bolts	43 "	
Nut on main shaft	22 "	
Clamp screw for shifting fork	18 "	
	22 "	(Synchromesh)
Rear axle nut	260 "	
Bolt for radius arm	70 "	

PINION NUT Standard: Tighten to 110 ft/lbs., back-off and retighten to 50 ft/lbs.

Synchromesh: Tighten to 125 ft/lbs., back-off and retighten to 56 ft/lbs.

Calling All Porsches

stood that the pump stroke controls the amount of fuel that is sprayed. The stroke is set at the factory to deliver .7 to .9 ccs. of fuel with each depression of the accelerator. The *duration* of the spray is controlled by the pump jet. The spray must hit directly onto the throttle butterfly and must, under no circumstances, be directed onto the sides of the carburetor, the venturi or the central fuel riser tube. If necessary, the discharge tube should be carefully bent to achieve this.

After starting the engine again, the idling adjustment can be made as follows:

1) Turn in one of the throttle stop screws to keep the engine running at the normal idling speed of 900 to 1000 rpm for the Types 527 and 528, and 600 to 700 rpm for all others.
2) Turn in the idling mixture volume screw of one carburetor gradually until the engine begins to falter and then back off approx. ⅛ turn. Do the same on the remaining carburetor. If the idling speed of the engine changes during this operation, it must be brought back to the correct rpm by adjusting the throttle stop screw. This operation may have to be repeated several times until the smoothest possible slow running is achieved. The second throttle stop screw must then be turned in until it reaches its stop.

This concludes the general engine une-up.

The Type 528 engine is frequently used for racing. It is fitted to the 1500 Super models as well as to the Competition Roadster. Proper tuning for racing will almost always spoil an engine somewhat for touring. In an air-cooled engine this condition is aggravated by slightly richer carburetor settings that are needed for engine cooling. In order to avoid excessive build-up of carbon, fouling of plugs etc., it is advisable to revert to standard tuning specifications as soon as possible after competing in racing events.

Engine preparation for utmost power output must include:

1) Fitting of colder spark plugs. The following plugs have been used successfully in racing the type 528 engine in the US: Bosch W 260 T-1; KLG F-100; Champion LA-11 and Lodge R-47. Warm up the engine on touring plugs and only install the colder ones just prior to the race.
2) Advance of the static ignition timing by 1/16 inch to augment the total spark lead at high engine rpm. At increasing engine speeds the automatic spark advance beyond the static ignition timing is approximately:
   2 degrees at 600 rpm.
   13 degrees at 1400 rpm.
   30 degrees at 2800 rpm.
3) Selection of correct carburetor settings. These will always have to depend on atmospheric conditions that prevail at the time and locality of the race, on the nature of the course and on driving methods. Theoretically speaking, if peak efficiency is required from the engine over a rather narrow range—say from 4500 to 6000 rpm—a larger venturi will usually be more effective than if good power is also needed far below these engine speeds. At high temperatures, high humidity and/or low barometric pressures the breathing capacity of the engine falls off and smaller fuel passages (main and pump jets) may have to be installed for a given venturi size. *Extensive tests are necessary to establish the best settings for each individual race meet. There is no short-cut to this procedure.* Whilst professional racing teams carry charts that have been assembled from previous experiences to determine a starting setting for final test runs, this material is not available to the owner-driver who competes in sports car events.

FROM the practical point of view, therefore, Porsche has established a *basic* setting to serve this purp e. It calls for a 29 mm venturi for all-round racing conditions. Carburetion test should be made with the settings that are given in parentheses on the carburetor chart, (Table B) and in accordance with the *Serial Number* where supplied.

Settings for two phases of carburetion must be determined by these tests; i.e. suitable fuel-air mixture for full throttle and for acceleration. The former is governed by a combination of the main jet and air correction jet; the latter by the main and pump jets. The main jet size must be established first. After driving at a steady half to three-quarter throttle under load for the better part of a mile, in a suitable gear for the prevailing road conditions, the ignition should be switched off and the gear lever placed in neutral simultaneously. A look at the spark plugs will reveal whether the mixture is correct. It must be assumed here that people who indulge in motor racing are familiar with spark plug "pictures." However, if the plug body at the end has a bright black appearance the mixture is usually correct; if sooty, the mixture is rich, or if dry grey in color, the mixture is too lean and a larger jet is needed. Main jets are available in .0005 graduations with the special additions of the 0107.5 and 0117.5 sizes.

After finding the proper main jet by this trial and error method, test runs must now be taken at full throttle, again under load. A lean mixture here calls for a smaller correction jet (located directly in the center of the carburetor throat) and a rich mixture for a larger one. Since most racing is done on full throttle, one must err on the rich side in this particular phase to assist engine cooling. Whilst a slightly rich mixture does not necessarily produce the highest possible car speed, it is definitely conducive to power and safety from break-downs in long races.

A few sudden acceleration tests from 3000 rpm will show up any faults in the mixture at this point. Hesitation without backfire through the carburetors will indicate the need for a smaller pump jet, while a larger one must be fitted if backfire occurs. Lastly, the idling mixture must be adjusted.

Air cleaners should be left in place for racing to reduce fire hazards. Their intake area is more than ample for the breathing capacity of the engine.

For any major service work on the Porsche engine, this unit is best removed from the car. Because of its compactness and light weight (approx. 170 lbs.) this can be accomplished in a fraction of the time required for conventional engines.

After removing the horizontal panel in back of the engine, disconnecting the main fuel line, the carburetor controls, the electric wires from the generator and coil and the heater cables, only two nuts at the bottom and two bolts at the top of the bell housing hold the engine in place. The chassis must now be raised and blocked. A garage type jack should then be placed under the crankcase and after removing the aforementioned nuts and bolts the engine can now be walked back and down, out of the car.

In the days of the long stroke-small bore engine, piston speed and related stresses were generally held to be the limiting factors of safe engine speeds. It has often been said of a certain type Bugatti that it can be driven for hours on end at the exact rpms that Le Patron laid down as maximum; but exceed this speed only slightly for a few minutes and expensive noises will be heard in the region of the crankshaft bearings. Exaggerated, I am certain. Be this as it may, since the advent of the ultra short stroke-large bore engine, these factors have become negligible and, except when overhead camshafts are fitted, the *valve gear* has become the limiting component. The proximity of the exhaust valve to the piston is bound to be much closer in a short stroke engine than in a long stroke one of equal compression ratio. Valve spring surge will usually set in and will give ample warning that the safe rev limit has been exceeded, before any damage is done to either piston or valve. In the heat of racing, however, one cannot always depend on "feel" and the tachometer must be watched.

The safe maximum rpm of the Porsche engine (that at which there is an ample safety margin) is 5500. To retain this margin, certain precautions must be exercised when reassembling cylinder heads after servicing. The valve springs should be tested for tension before installation. The inner spring should have 14 lbs. when compressed to 1 15/32-in. and 33 lbs. at 1 5/32-in. The outer one, 33 lbs. at 1 17/32-in. and 84 lbs. at 1¼-in. The total length of the installed spring should be 1 19/32-in. on the Type 546 (America) and 1 17/32 in. on all others. This can be adjusted by the removal or addition of shims under the valve spring. Valve rockers must swing through the correct arc. At half valve lift the rocker must be at 90 degrees to its push rod. To achieve this, the rocker shaft may have to be raised by shims. Care should also be taken that the exhaust rockers do not foul the valve spring retainer plates, and if necessary some material may have to be ground away. At this point study Table D.

THE Type 369 Porsche engine uses cast iron cylinders, all others are fitted with light alloy ones that have the chromium bore plated directly onto the aluminum. Because of the very close and exact tolerances that are necessary in the latter, pistons are not readily replaceable and both piston and cylinder must be changed in the rare cases where this need arises. If time is not of the essence, however, matching pistons for existing, serviceable cylinders can be ordered from the factory. In such case the *complete code number* that is stamped on the *base* of each cylinder and on the *crown* of each piston *must be supplied with the order.*

Wrist pins have a palm push fit in the piston bosses, at 200 degrees Fahrenheit. They should, therefore, never be driven out cold, or bending of the conrod may result. Instead, pistons should be heated evenly by playing a flame *over the crown only,* to avoid distortion, and the pins pushed out with a suitable drift.

Experience has taught that piston rings should be renewed whenever a cylinder is removed from a piston. The rings must be fitted individually to the bore and the proper gaps are: for the Type 369—.015-in.; all others—.004-in. The piston skirt clearance is .0015-in.

The crankshafts in all Porsche engines run in four replaceable main bearings of aluminum alloy. The main bearing clearance is .002 to .004-in. (when the bearings have been compressed by the bolting together of the crankcase halves). Con-rods are fitted with caged roller bearings in Type 502, 527 and 528 engines, and with replaceable lead-bronze plain bearings in all others. The latter have a clearance of .001 to .003-in.

Engine lubrication is by gear pump and, with the oil at operating temperature, the pressure varies from seven psi at idling to 35 psi from 2500 rpm up.

The fuel pump is of the mechanical, diaphragm type with a pressure of from 1.5 to two psi.

To avoid confusion which often arises in seeking a burned fuse (the fuse box does not carry a diagram), the arrangement of the fuses in the box, reading from right to left, is as follows:

1—Stop light, directional signals; 2—Cigar lighter; 3—Horn; 4—Windshield wipers, interior (dome) light; 5—Dash board lights; 6—Right tail light; 7—Left tail light, license plate light; 8—Parking lights; 9—Low beam of right head light; 10—Low beam of left head light; 11—High beam of right head light, and blue control light on dash; 12—High beam of left head light. All fuses are of 8/15 Amps capacity with the exception of the one for the cigar lighter which is 25/40 Amps.

The Models America (with Type 546 engine) and Super (with Type 528 engine) are fitted with the new Porsche synchro-mesh transmission. The method of synchronization is an entirely new conception in this direction and another Porsche patent. While this transmission lends itself to extremely fast shifting, one must never forget that stresses placed upon synchronizing parts *increase with the square of their rotating speeds.* Down-shifts, therefore, should be confined to strict maxima of:

75 mph from 4th to 3rd
35 mph from 3rd to 2nd
10 mph from 2nd to 1st

When indulging in racing downshifts at higher speeds one must treat the transmissions exactly as one of the crash-types and *always double clutch.*

When filling the transmission with lubricant, the level should *never* be above the bottom of the filler neck or leakage through the rear wheel grease seals will result.

Cars that are kept in the open, or that are constantly driven over smooth roads only tend to lose suppleness of the front suspension. Frequent lubrication of the trailing arm bosses with SAE 90 gear oil will prevent this.

Tire pressures are a matter of personal preference, but rear pressures should exceed front pressures by from 25 to 33⅓ per cent. For comfort, 19 psi front and 23 psi rear constitute the absolute minimum. Where lateral stability is sought, 23 psi front and 30 psi rear have been successful for racing.

During the last two racing seasons Porsche cars have literally swept the boards. It has become a byword that "a properly prepared and properly driven Porsche can only be beaten by another Porsche in the Class F Production Car Class." Although not every Porsche owner uses his car for racing, pride of ownership will compel him to preserve the performance with which his car was originally endowed. In closing, therefore, I would strongly advise periodical visits to your Porsche dealer for inspection and service, but for those owners who are amateur mechanics, Torque Wrench specifications in Table D may be helpful.

No. 1513: PORSCHE TYPE 356 SALOON

The fine aerodynamically inspired lines are apparent, in spite of being somewhat marred at their beginnings by massive over-riders. The handle for raising the front locker lid takes the form of a simple motif, and the door locks are operated by push buttons.

The Autocar ROAD TESTS

THE late Dr. Porsche, the brain behind the pre-war Auto-Union racing cars, was a man of very advanced and sound engineering ideas. The Volkswagen was developed from his design, and the Porsche car which is being manufactured at Stuttgart in the Western Zone of Germany. The Type 356 Porsche which is the subject of this Road Test, and which was provided by Colborne Garage, Ltd., of Ripley, Surrey, the concessionaires in Britain, is a most interesting 1½-litre car both from the constructional angle and in the way it behaves on the road.

In brief, it is of chassisless construction, the suspension being independent both front and rear. The flat four-cylinder air-cooled engine is mounted at the rear and is coupled to a four-speed all-synchromesh gear box by a single-plate clutch.

By virtue of its very low build and fine aerodynamic lines it attracts immediate attention and interest from young and old. It is so obviously a car that was designed by men who knew what they wanted and were able to carry out their ideas. Its very appearance suggests speed, and as soon as one is seated in the car any desire to loiter is quickly dispelled. There is no doubt that low weight allied with effective streamlining enables a high performance to be achieved by a comparatively small engine. As always, the ability to travel quickly is useless unless the car in question is able to hold the road. The Porsche does this in no uncertain manner, the soft torsion bar springing allowing it to hurry round main road corners without roll, while the rather direct steering gives the driver exact control over the front wheels.

A greater number of miles than usual were covered during the test and the more experience one gained the more enjoyment there was. The high top gear makes cruising effortless with an indicated 75-80 on the speedometer. Long trunk road gradients can be surmounted on top, but the gear box must be used to obtain the best performance, and one can imagine the car being thoroughly at home storming Alpine passes (as its success in this year's Alpine Rally demonstrated forcibly), where the admirable third gear and also second could be used to advantage. At night there is the impression of being in an aircraft cockpit with the close curved windscreen and discreet lighting from the facia, the suspension ironing out any sudden undulations in the road surface and no squeal being evident from the tyres when cornering fast. There is a feeling of rushing through space with the road disappearing rapidly immediately in front and the subdued beat of the engine from the rear.

The air-cooled engine builds up its revs quickly in the indirect and while a maximum reading of 91 m.p.h. on top gears was obtained in the available distance, on the electric speedometer used for test purposes, there were indications that, given a longer run, the car would exceed this figure by a comfortable margin. When the engine is turning over quickly there is some noise from the belt-driven cooling fan, although this is diminished if the windows are closed. Wind noises, as might be expected from the shape of the body, are almost non-existent and the effectiveness of the streamlining is proved by the speeds obtained during the test, when there were very adverse wind conditions. Occasionally, at about 60-70 m.p.h., an intermittent slight thudding was felt on the eardrums when the windows were fully open; this curious effect could be eliminated by closing the windows and opening the rear quarter lights.

The gear change, operated by a short, slender central lever with a large knob, is one of the most pleasant and certainly one of the fastest manually operated changes experienced. It is possible to make noiseless changes just as fast

Left: The battery is located behind and below the spare wheel. The tool kit includes a graduated wooden dipstick for the fuel tank, which has a large bayonet-action filler cap.

Right: The two occasional rear seats are reasonably easy to enter with the back rests of the front seats tilted forward.

Louvred wheel discs help to cool the brake drums and are in keeping with the general appearance of the car. The rear quarter lights open sufficiently at the rearward edge to provide ventilation inside the car with the front windows closed. The engine cooling fan draws air through the plated grille below the rear window. The small round cap in front of the rear wheel covers the adjustment of the rear suspension. Right : A striking impression of the frontal shape is given in this view. Flashing-type direction indicators are combined with the side lamps. The bumper bar wraps round the corners of the car and has a central rubber moulding.

as the driver can move the lever. The actual movement from gear to gear is greater than on some cars with a central lever. To achieve gear changes such as described the clutch pedal must be depressed to its fullest extent. This is no hardship, as the operation is light and there is no trace of judder in the clutch engagement.

The worm geared steering, with a divided track rod, gives the driver a feeling of definite contact with the front wheels. There is pronounced oversteer, as with all rear-engined cars, and, as with a racehorse which might bolt unless a firm hand is on the reins, so must one be in control here, as it is possible to bring the tail round very quickly if the driver is too enterprising on wet surfaces. The suspension is soft and fast cornering can be indulged in without sign of roll. Rough surfaces make little difference to the comfort of the occupants, although there is some hammering from the small-diameter wheels.

Brake Behaviour

The brakes are in keeping with the performance, the car pulling up straight even with all wheels locked. The pedal pressure required for an emergency stop is fairly heavy, but at no time was any tendency to fade noticed. The hand brake lever is tucked away under the facia and is a little awkward to find at times. The controls are well placed, the hands falling naturally on the steering wheel. The brake and throttle pedals are so placed that "heel and toe" changes are convenient, while there is room for the left foot to clear the clutch pedal.

Because of the low build, getting in and out calls for a certain agility—a knack is acquired—but the driving position is very good, the view through the one-piece curved screen being unobstructed, whilst rearward vision in the driving mirror is satisfactory. At night head lights from following vehicles have an unpleasant blinding effect in the mirror unless it is moved out of focus. Passengers in the front seat commented on the unusual effect caused by lack of the more normal view of bonnet and radiator. Separate front seats, upholstered in fine-quality leather, are most comfortable, no fatigue being apparent after a journey on which a high average speed was achieved. The back rests are adjustable for angle by means of screws and lock nuts, and the fore-and-aft adjustment is generous, allowing a position to be obtained by drivers of varying height which gives a maximum feeling of control. The rear compartment has two emergency seats, suitable for two children or two small adults. The back-rest angle, again, can be varied. When folded flat the back rest forms the floor of a compartment for carrying luggage.

The facia, in which there is a glove box, has provision for a radio to be fitted with the speaker in the centre of the panel. In addition to the speedometer there is a rev counter, the needle of which on the car tested was apt to float a great deal. The only other instrument is an oil temperature gauge. There is an oil warning light, but no fuel tank gauge. A reserve of one gallon of fuel is available, the lever operating the change-over tap being in the centre of the bulkhead; the tap itself is close to the footboard and is apt to be touched by the passenger's feet. Two notes for the horn are provided by a change-over switch.

Lighting of the facia panel is effective and unobtrusive. An annoying reflection occurs during daylight in the concave glasses of the speedometer and rev counter. Winking direction indicators are fitted, operated by a neat arm below the steering wheel, and there is a small map pocket in the carpet type lining on each side of the front compartment. An interior light comes on when either door is opened. The main head-lamp beam could be more powerful to be in keeping with the performance. The double-dipped position with its flat top cut-off, actuated by a convenient foot switch, aroused no opposition from oncoming traffic.

The heating and demisting equipment is one of the most effective encountered, really hot air being blown by the

The air-cooled flat-four engine is located above and behind the final drive. Cooling is achieved by means of an enclosed fan, which is driven off the rear end of the dynamo. Auxiliaries are accessible, the oil filler being to the right of the dynamo.

An air of comfort and spaciousness is apparent in the front of the Porsche, in spite of its modest overall size and low build. The flat rubber-covered floor is well below the level of the door sills.

engine cooling fan along ducts to the side of the body just in front of the seats and also on to the windscreen through slots in the facia. The supply of hot air can be adjusted by means of shutters over the outlet ports alongside the seats. At times, when using the maximum heat position, slight traces of fumes were noticeable inside the car, with all the windows closed. Cold air can also be circulated through the same ports by operating controls just below the facia sides. A windscreen washer was fitted as an extra, being operated by a single-stroke plunger below the facia.

The spare wheel, battery, tool kit and the fuel tank, with its very large filler, are all contained in the forward locker. There is also space there for small items of luggage, the floor of the locker being lined with rubber matting, as is the floor of the car itself. The locker lid, released by a control knob beneath the facia, is held open automatically when raised. There is an emergency method of releasing the lock. There are 16 grease nipples which require attention approximately every 500 miles.

The engine fits snugly into the rear compartment, the lid of which is released by a control knob behind the left-hand front seat. The oil filler is accessible, as is the distributor, but two of the four sparking plugs are rather tucked away behind the inlet manifolds. The engine started quickly from cold without use of the choke and gave no sign of pinking on first-grade fuel.

For the driver who wants a car which is different, albeit expensive, with import duty and British purchase tax added, the Porsche has a definite appeal. It is fast and comfortable, is well made, and the general finish of the fittings, upholstery and paintwork is an example of how thorough the German can be. It steers well, holds the road in a manner of its own, and by reason of its shape and light weight it is economical.

PORSCHE TYPE 356 SALOON

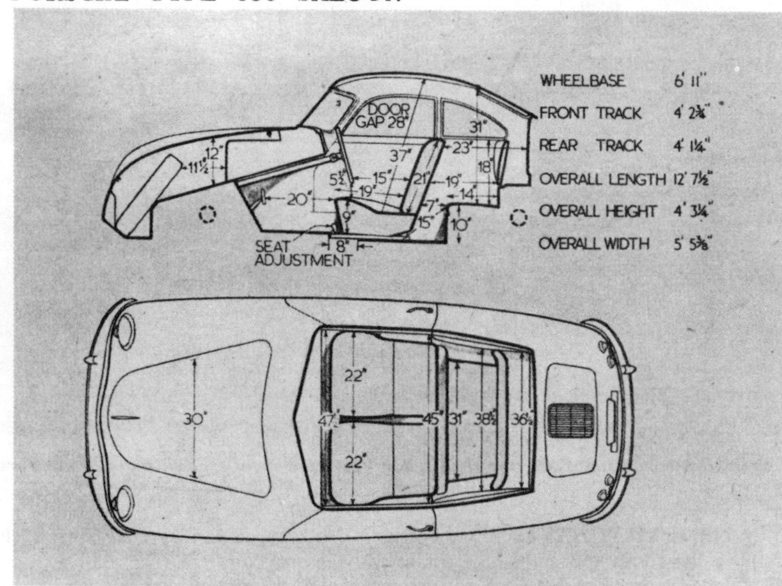

Measurements in these ½in to 1ft scale body diagrams are taken with the driving seat in the central position of fore and aft adjustment and with the seat cushions uncompressed.

PERFORMANCE

ACCELERATION: from constant speeds. Speed, Gear Ratios and time in sec.

M.P.H.	3.54 to 1	4.95 to 1	7.9 to 1	13.9 to 1
10—30	—	8.9	5.5	—
20—40	13.1	8.1	5.7	—
30—50	13.2	8.4	6.4	—
40—60	14.4	9.7	—	—
50—70	15.4	11.0	—	—

From rest through gears to:

M.P.H.	sec.
30	5.1
50	10.6
60	17.0
70	21.2
80	31.9

Standing quarter mile, 20.1 sec.

SPEED ON GEARS:

Gear		M.P.H. (normal and max.)	K.P.H. (normal and max.)
Top	**(mean)**	87	140
	(best)	91	146
3rd	..	60—76	97—122
2nd	..	38—50	61—80
1st	..	18—27	29—43

TRACTIVE RESISTANCE: 38 lb per ton at 10 M.P.H.

TRACTIVE EFFORT:

	Pull (lb per ton)	Equivalent Gradient
Top	158	1 in 14
Third	240	1 in 9.3
Second	400	1 in 5.5

BRAKES:

Efficiency	Pedal Pressure (lb)
77 per cent	132
70 per cent	100
37 per cent	50

FUEL CONSUMPTION:
28 m.p.g. overall for 784 miles (10.1 litres per 100 km).
Approximate normal range 27-30 m.p.g. (10.5-9.4 litres per 100 km).
Fuel, First grade.

WEATHER: Dry; high wind.
Air temperature 60deg F.
Acceleration figures are the means of several runs in opposite directions.
Tractive effort and resistance obtained by Tapley meter.
Model described in *The Autocar* of December 12, 1952.

SPEEDOMETER CORRECTION: M.P.H.

Car speedometer	10	20	30	40	50	60	70	80	97
True speed	10	19	28	37	45	55	65	75	91

DATA

PRICE (basic), with two-door saloon body, £1,367 (including import duty).
British purchase tax, £604.
Total (in Great Britain), £1,971.
Extras: Radio £40. Heater standard.

ENGINE: Capacity: 1,488 c.c. (90.7 cu in).
Number of cylinders: 4.
Bore and stroke: 80×74 mm (3.14×2.9 in).
Valve gear: Overhead; push rods and rockers.
Compression ratio: 7 to 1.

B.H.P.: 55 at 4,400 r.p.m. (B.H.P. per ton laden 57.8).
Torque 78 lb ft at 3,200 r.p.m.
M.P.H. per 1,000 r.p.m. on top gear, 23.

WEIGHT (with 5 gals fuel), 16 cwt (1,792 lb).
Weight distribution (per cent) 44.7 F; 55.3 R.
Laden as tested: 19 cwt (2,128 lb).
Lb per c.c. (laden): 1.4.

BRAKES: Type: F, Two-leading shoe. R, Leading and trailing.
Method of operation: F, Hydraulic. R, Hydraulic.
Drum dimensions: F, 11¼in diameter; 1¾in wide. R, 11¼in diameter; 1¾in wide.
Lining area: F, 78.7 sq in. R, 78.7 sq in (165.6 sq in per ton laden).

TYRES: 5.00—16 in.
Pressures (lb per sq in): 20 F; 25.5 R (normal).

TANK CAPACITY: 12 Imperial gallons (including 1 gallon reserve).
Oil sump, 5 pints.
Cooling system, air.

TURNING CIRCLE: 34ft 0in (L and R).
Steering wheel turns (lock to lock): 2½.

DIMENSIONS: Wheelbase 6ft 11in.
Track: F, 4ft 2¾in; R, 4ft 1¼in.
Length (overall): 12 ft 7½in.
Height: 4ft 3¼in.
Width: 5ft 5⅜in.
Ground clearance: 6¼in.
Frontal area: 17.3 sq ft (approximately).

ELECTRICAL SYSTEM: 6-volt; 75 ampère-hour battery.
Head lights: Double dip; 36—36 watt.

SUSPENSION: Front, by independent double arms with laminated torsion bars.
Rear, swing axles with torsion bars.

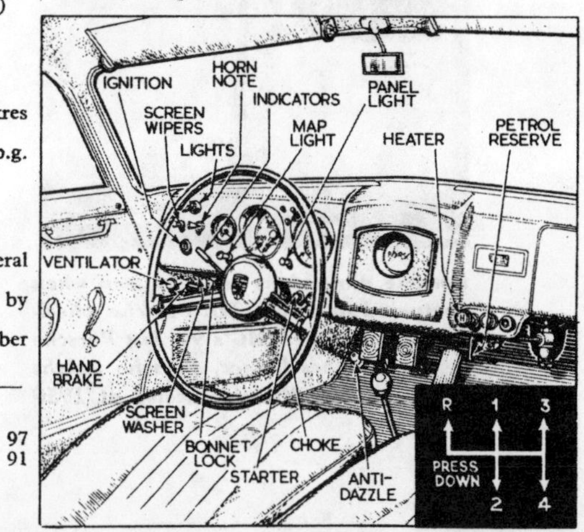

BECAUSE IT'S A PORSCHE

The author (rear car) and a friend put their fast-cornering Porsches through a fast curve on a country road in Germany

...It's built with care belying a life span of just five years

By T. C. Countryman

Porsche dimensions are best seen when compared to an MG-TD. Porsche wheelbase is 83 inches, MG's 94, but Porsche is seven inches longer. Porsche weighs 1640 pounds, MG tips scales at 1950

A NAME THAT HAS LIVED to become a legend in the automotive industry now adorns the hood of one of the most radical, superb automobiles in sports car annals. This small, rear-engined car was designed by the late Dr. Ferdinand Porsche, an automotive pioneer, and a guiding light behind such bright standouts as the Astro Daimler, the SSK Mercedes and the rear-engined Auto-Unions that dominated pre-war Grand Prix racing in Europe.

The present-day Porsche's ancestry can be traced back to one of Dr. Porsche's life-long dreams — a mass-produced small car that would be within financial reach of the entire German populace. This car, utilized as a political toy by Adolph Hitler, became known as the Volkswagen — the People's Car. With the coming of World War II, the project was laid aside. But a start had been made; the German government produced a military Volkswagen — a German Jeep. This vehicle had innate similarities to today's VW: specifically, a box-like, pressed-steel frame, and a horizontally opposed, four-cylinder, air-cooled engine. After the war, and with the aid of Marshall Plan money, the VW became a reality, and today there are nearly as many Volkswagen cars as there are bicycles on German streets.

With the Volkswagen a very real thing, the next step was obvious, and eagerly anticipated by Porsche's followers — a machine of perfection, similar in many ways to the "People's Car," but with a world of difference in finesse and application between the two cars.

It was while he was in prison (sentenced there by an Allied tribunal for his part in wartime production) that Dr. Porsche perfected his plans for the car that was to bear his name.

The first Porsche was built in 1949 in a small town in Austria. After only a few of these handmade cars were produced, Dr. Porsche and his small staff of engineers moved into the Reutter Body Shop in Zuffenhausen, a suburb of industrial Stuttgart. The new location was just across the street from the original Porsche plant, which had been taken over by Allied occupation forces for use as an ordnance shop.

Although a new building has been built to house the Porsche staff and its limited assembly plant, all major Porsche components are made by other firms; Reutter of course carries the contract for the bodies. Porsche assembles only the engines and

Photos by T. C. Countryman

running gear and places them in the Reutter bodies. The major role played by the Porsche staff is that of a firm of consultant engineers; of the 350 Porsche employees, over 50 of them are engineers, designers of machinery for other companies.

The Porsche production line (a rather eloquent phrase, comparing its eight-a-day output to the larger factories of Europe and the U.S.) is limited to three body styles: the familiar coupe, a trim convertible, and on special order, the sport roadster. At present, four engine sizes are available. (Although I say "at present," it's believed that by the time this issue goes to press, the new Porsche 1.5-liter "550" will be included in general production, meaning an additional body type — a high-fendered, aluminum-bodied roadster — and another mechanical masterpiece, the 550's quad-cam, 110-bhp engine.)

The lowest-output Porsche engine, a 1.1-liter job (67.1 cu. in.), develops 40 horsepower. (This engine is a mate to the Volkswagen powerplant in displacement, but the Porsche touch brings out 14 more horses.) The next engine, following a demand for increased torque, is the 1.3-liter model; it puts out 44 horsepower. This isn't quite to the liking of those who want to challenge the 1500 cc competition; thus the lively 1.5-liter Porche engine is on the list. A natural for racing, the 1500 cc powerplant has a flexible 55-horsepower output. (Early 1.5 engines boasted 60 bhp.)

For performance-plus, the 1.5-Super model puts out a lusty 70 bhp — a must for those who would rather race than eat (and you can take that literally, considering the over-$4500 price tag on the Super).

Take a 165-pound, 70-horsepower engine, put it into a very streamlined car weighing about 1640 pounds (dry), and

CONTINUED ON PAGE 12

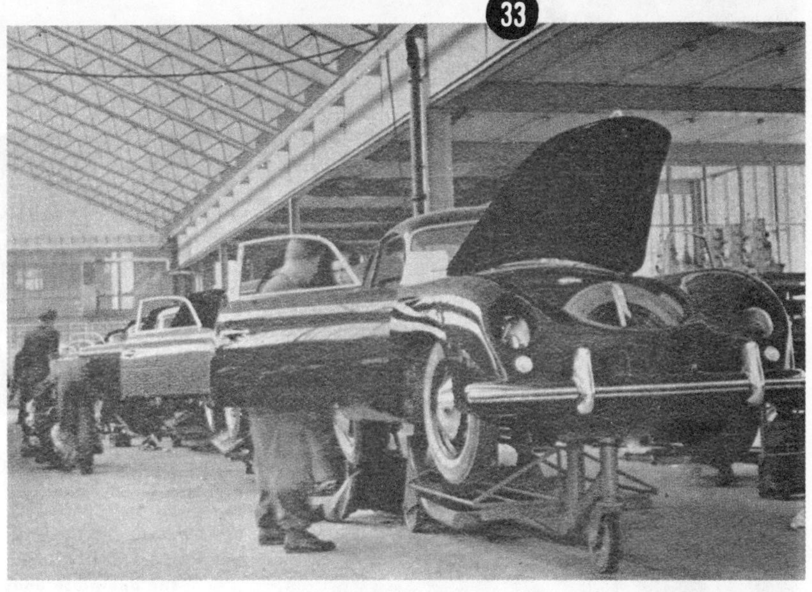

Normal production line output is about seven or eight cars a day. When completed, this car will be shrouded to protect its finish, then road-tested on Autobahn

Newly assembled engines are flushed with oil to remove foreign particles, then run two hours at 1000 rpm. After engine assembly, owner's name is chalked on muffler

Engine assembly is a painstaking, exacting job. Here, a Porsche craftsman weighs connecting rods for perfect match

One of eight Porsche master mechanics on the engine line checks ring clearance in chrome-plated aluminum cylinder

Exacting labor goes into the Porsche, one of the few cars in the world made chiefly by hand

PHOTOS BY WIDE WORLD

The one-piece body of the Porsche car sits in the midst of its main parts, which are chiefly handmade. Engine, located in rear, is at left in front.

Handmade Handy Car

Hours of skilled work have been expended on this Porsche, one of the few cars in the world made chiefly by hand. The cars, all two-seaters, are produced in Stuttgart, Germany, and are capable of speeds up to 109 mph., depending on type of engine.

An average day's output on the line at Porsche factory in Stuttgart. The compartment at rear is not for luggage, but is the place for the engine. Gas tank is in front.

Like a craftsman of the Middle Ages, a mechanic puts the finishing touch to a Porsche engine. Chiseling his name on motor block, he takes full responsibility for the job.

In the Porsche factory at Stuttgart, skilled craftsmen put together the machines. At left is the plant's only assembly line. The rear-engined auto was designed by the late Ferdinand Porsche, and his son, Ferry, is now the head of the company.

Years Ahead in Engineering

May 1955

The marks of a thoroughbred are seen at a glance. The dashing lines of every PORSCHE immediately satisfy the critical eye, give dramatic promise of thoroughbred road performance. And that promise is fulfilled beyond your dreams by PORSCHE race-tested engineering: famous PORSCHE-developed synchro-ring transmission, torsion bar springing, brakes twice normal size for the car's overall weight. These are a few of the years-ahead features that make PORSCHE a consistent class winner in all major international races and rallies. PORSCHE puts you miles ahead in driving thrills and safety on any road in the world!

Every PORSCHE today bears the stamp of this man's creative genius. One of the giants of the Automotive Age, Prof. Dr. Ing. h. c. F. Porsche devoted 50 years to developing the basic engineering principles that continue to make PORSCHE the world's leading sports car, year after year.

Miles Ahead on the Road

Speedster

. . . the sports car that's *all* sports car, developed from the long line of PORSCHE racing champions. Brilliantly streamlined, completely functional in concept, every inch designed for your maximum pleasure on the wide open road. Top speed: 100 miles per hour.

$2995, delivered N. Y. C.

PORSCHE MOTORING

Some Reflections After 30,000 Miles With a "Damen"

SMOOTH.—The outline of the standard Type 356 Porsche saloon is very pleasing to the eye in spite of the absence of conventional radiator. Its air-flow is satisfying to the most finicky aeronautical engineer, though the weight of 17½ cwt. is depressing. Even so 55 b.h.p. with a good torque curve is sufficent to give it a fair performance for a 1½-litre two-seater saloon.

BEFORE becoming too deeply involved in the question of driving a Porsche it might be as well to pay a quick visit to the factory and find out something about the cars. Situated on the north-west edge of Stuttgart, a mere three minutes from an autobahn, this small factory is entirely new, the old one having been damaged and requisitioned by military authorities. It was in this original factory that Dr. Ferdinand Porsche had his design and development business, the new factory being built when the construction of Porsche cars began in 1950. Since the death of Dr. Porsche, who was a pure inventor as well as a designer, the firm has been effectively carried on by his son Dr. Ferry Porsche. Apart from car manufacturing, the Porsche concern still do contract design work for all branches of the engineering industry and among the far-reaching effects of the Porsche genius are the new gearboxes on the Grand Prix Bugatti, and experimental B.M.W.s, both of which are built under Porsche patents.

The factory is divided into four departments: machine shop, assembly, service and racing, the last dealing exclusively with competition motoring, the building of the sports/racing "Spyder," the assembly of the 4 o.h.c. Carrera engines and experimental work. When I went to take delivery of my Porsche coupé I took the opportunity of having a look to see how it was made and when I eventually drove away I had already gained a certain amount of confidence in the car. The Porsche is very definitely a hand-built car, the output being in the order of six or seven cars per day, the assembly line running from the front of the factory to the rear, so that when you arrive in the morning you might see a green coupé joining the line and by the end of the day it will be near the end about to receive final adjustments, a continuous flow of about nine or ten cars being on the assembly line. Next door to the Porsche factory is one belonging to Reutter the coachbuilders, and the two are joined by a short private road. This branch of Reutter is occupied solely with Porsche work, and while it is not owned by Porsche it is under their control as far as work is concerned.

There is no chassis in the normal sense of the word, but a flat platform with box-sectioned sides and box structures at the front end and at the scuttle. The whole thing is made of sheel steel pressings which are spot-welded together and it was interesting to see that in some places where the sheet was bent without a former the ripples formed on the surface of the metal were left as they had the same stiffening effect as an intentional corrugation or a pressed-out stiffening channel. At the front are two tubular cross-members to carry the suspension units, another at the rear and a large one across the scuttle joining the rectangular section uprights on which the doors are hung, while these members are used to transmit hot air to the screen ducts. The bodywork, whether the fixed-head coupé, the drophead or the open Speedster, is also of sheet steel and when built is fitted to the platform chassis and welded into place, thus forming a very strong and rigid monocoque. The body is then trimmed, glazed and painted and "undersealed," but as yet there are no mechanical components on the car, but as the platform was built on a one-piece jig all the necessary holes to take bolts, bushes or bearings are finished. This complete body/chassis unit is then taken into the Porsche factory and placed on a three-wheeled trolley, so that it is about three feet off the ground, and it then joins the assembly line. On the base of the trolley are trays containing all the smaller mechanical parts for the chassis, while larger components, such as seats, steering columns and so on, are on racks alongside the assembly line at the points at which they will be required.

Most of the mechanical components, in the shape of castings, forgings or stampings, are made by outside firms to Porsche specifications and drawings and subject to Porsche inspection before being used, while all machining requiring great accuracy is done at the factory. Engines are assembled by individual fitters, each one being responsible for the whole unit, and the rear axle/gearbox unit is built in the same way. The two major components are assembled on either side of the main line and fitted to the car as complete units. The completed car is removed from its trolley and placed on a special machine on which wheel alignment, castor angles, camber

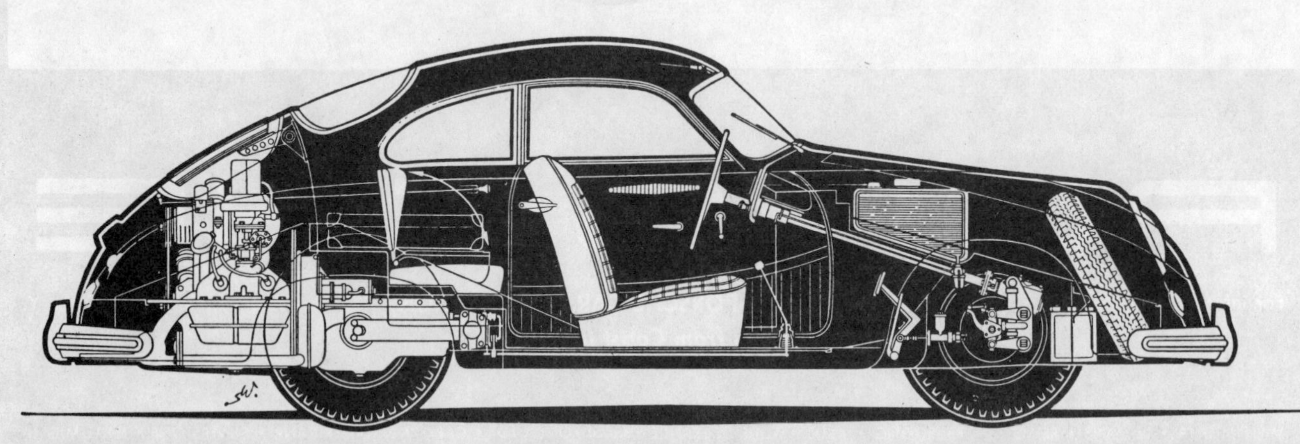

INTERNALS.—The disposition of the various components are clearly shown in this sectioned drawing of the Porsche 356. At the front are 11-gallon fuel tank, spare wheel, tools and battery, while behind the seats is space for two or three suitcases, squashy bags, briefcase and a vast quantity of odds and ends. Alternatively it will take two children or a small adult, while three people in the front is not uncomfortable.

angles, toe-in, etc., are all checked to micrometer accuracy and the car is then given a 60-mile road-test after which any further adjustments required are effected and it is then ready to leave the factory.

There are three basic types of engines built, all having four horizontally opposed air-cooled cylinders, but the main production is concentrated on the normal or " Damen " (the lady) and the Super. These two differ in compression ratio, valve and port sizes, exhaust system and crankshaft, the normal have conventional rods and lead-bronze big-ends and the Super have roller-bearing big-ends with non-splittable big-end eyes in the con-rods, the whole assembly being built-up by the Hirth Company on their patent splined journals and webs. These Super cranks and rods arrive at the factory already assembled, the others being machined and assembled at Porsche. The inlet valves on the normal engine are vertical and the exhaust valves inclined, both operated by pushrod and rockers, a single camshaft centrally placed above the crankshaft having four cams on it which operate the eight valves, each cam working opposed valves. The whole principle of the engine is like the VW, which is not surprising as Dr. Ferdinand Porsche designed that particular vehicle, and the air-cooling is effected by a large fan on top of the engine and driven off the dynamo. This fan blows air through ducts to close-fitted shrouds around the cylinders and heads, the hot air escaping underneath the car. In the main duct an oil cooler is mounted so that the oil is cooled independently of the speed of the car, which is very useful when storming a mountain pass in second gear, for no matter what the road speed the faster the engine runs the more cooling there is to the oil; this also applies in traffic driving. This same layout is followed on the 1½-litre " Spyder " engine, which is now giving 115 b.h.p. at 6,200 r.p.m., but on this unit there are two overhead camshafts to each bank of cylinders. A roller-bearing crank is fitted and from the clutch end there is a gear-drive to a short shaft underneath the crankshaft; this shaft drives two bevel wheels mounted back to back, each in turn driving another bevel gear coupled to another shaft so that the rotation of the main shaft is turned through a right angle in each direction, the secondary shafts running one between each pair of cylinders. When these shafts, enclosed in tubes, reach the cylinder head they drive directly onto a bevel gear mounted in the centre of the exhaust camshaft, which is the lower one on each side of the engine. From this camshaft bevel gear another bevel and short shaft runs upwards to a similar pair on the inlet, or upper, camshaft, this system giving an inclined valve layout in each cylinder. A double-choke carburetter feeds to each pair of cylinders, whereas on the push-rod engines a single-choke carburetter is used, though it is possible to fit double-choke instruments. Twin plugs per cylinder are used on the " Spyder " engine, against single plugs on the normal and Super, and naturally the camshaft engine has a larger cooling fan, still driven by belt with the dynamo mounted on the fan shaft, the whole engine being closely ducted by metal shields. This unit differs from the normal engines by having dry-sump lubrication, with a separate tank and oil cooler in the nose of the car on the racing models.

The production version of this camshaft unit has a slightly lower compression ratio and is known as the " Carrera " engine, so named since its proving-ground was the Carrera-Mexicana or Pan-American road race; Le Mans and Mille Miglia also played a great part in proving the reliability of this engine. Competition is a by-word at the Porsche factory and the Porsche car is a perfect example of a race-bred one, the 1956 models all being available with the Carrera engine. Until the recent Frankfurt Motor Show the push-rod Porsches were sold in 1,100, 1,300 or 1,500-c.c. form, either "Damen" or " Super," the only differences being in the bore and stroke, the rest of the car being the same for all models. Now, however, the 1,100 has been dropped and the 1,500 has been enlarged to 1,600 c.c., that being the International capacity dividing line for Gran Turismo cars, and both sizes are still available in normal or Super form. If you want a car for touring purposes, Porsche recommend the normal engine, the Super being a competition version for rallies or races, and both units being interchangeable a keen owner could have one car and two engines, a " cooking " one and an " eating " one. The " eating " engine being intended for competition, its life is not anticipated at more than 40,000 miles, whereas the " cooking " one is reckoned to do 60,000 miles before worrying about wear and tear. As my use of the car was to tour Europe in reach of motor races I had a 1,500-c.c. normal, " a nice car," said Porsches, " that is reliable and long-living, but rather dull, it does only 95 m.p.h. and 5,200 r.p.m."

Before I made the choice of a Porsche I had to suffer a great deal of barracking with remarks such as " My God, the oversteer," or " they are impossible in the wet," " the fan belt breaks and then where are you?" " it will spin as soon as you see a corner," "you need a bath after changing the plugs," " they are noisy and rough," and " bloody clockwork contraption." I put up with all these because I thought the aerodynamic shape of the saloon looked about right and it was a car I could lean my elbow on (my friends mutter rude things to me about dwarfs when I discuss heights !). I have now completed 30,000 miles of Porsche motoring, all of it spelt with a capital M (see MOTOR SPORT for December, 1954, page 699) and the oversteer I enjoy having learned to drive on a chain-gang Frazer-Nash, the wet petrifies me in any car, the fan-belt is original and has not even needed adjusting, I have only spun it once in the total distance covered, the plugs I changed as a routine every 10,000 miles and the engine keeps so clean I only needed to wash my hands, which I reckon to do within 10,000 miles anyway; I never had occasion to look at a plug while on the road, this being 1955. The noise is outside the car, not in, and the roughness sounds on the Super model but not on the " Damen " and I found my model has a jolly strong spring in its " clockwork mechanism."

The whole essence of driving a Porsche lies in the fact that everything is finger-light ; the steering, clutch, gear-change and brakes are all of a light smooth feeling that at first comes as strange after conventional cars. It has a live feel in its manner of going that wants caressing, not taken firmly between clenched fists as on some cars, while this manner of going is something that the driver has to accustom himself to. If you approach a Porsche with a view to driving it like a conventional car you will hate it, but on the other hand if you are prepared to spend say 1,000 miles in learning to drive all over again, you will love it. The Porsche is essentially a sporting car and likes to be driven in a sporting fashion, in fact the harder the better, and you find after a time that there are a number of things about it that you must absorb into your system. One is an appreciation of the rev-counter, for while the engine pulls happily from 1,500-5,000 r.p.m., it pays dividends to keep it between 2,500-

SPYDER.—The 4 o.h.c. Porsche engine of 1½ litres is now in production to be fitted to any model. On the racing Spyder this unit is mounted in front of the rear axle and turned round the other way. The camshaft boxes, twin-choke carburetters and large cooling fan are visible on this Spyder unit.

CONTROL DEPARTMENT.—The two-spoke Porsche wheel is designed for 90 m.p.h. cruising with the finger tips resting lightly on the spokes, a practice recommended by Porsche. Instruments are clear and simple, being r.p.m. and m.p.h. with oil temperature in the centre. This is the rather spartan Speedster cockpit, with racing-type bucket seats.

4,500 r.p.m., and it doesn't wear out. Another is to realise that the direction in which the nose of the car is pointing is of no importance, providing the driver is convinced of the way he wants to go; and finally, before you start going quickly in a Porsche you must be able to *drive* anyway. I have met lots of people who have tried a Porsche and thought it terrible and when I have seen them driving a conventional car I understand why; they just cannot *drive* properly anyway. With its low build, trailing-link i.f.s. and swing-axle rear suspension and rear-mounted engine it has an obvious oversteer characteristic, but this is constant and progressive and not changeable and sudden. The worst thing is surely an understeering car that changes its characteristics to violent oversteer in the middle of a fast corner. With the Porsche there are three ways of taking a corner: first at touring speeds, when there is no roll at all and the steering is neutral; secondly at fast road speeds when there is still virtually no roll but a slight oversteer which requires you to unwind the steering slightly before you leave the corner, this unwinding being progressive and varying with radius of the corner. Thirdly, there is the method for *very* fast cornering and this is where the Porsche technique must be applied; if it is not then you find yourself in a classic vintage oversteering slide on full opposite lock and about to turn right round. This special Porsche technique is better described by the acknowledged German expert on this type of motoring Richard von Frankenberg and I quote from an article he wrote for the *Sports Car Club of America*. "Often it is said that the Porsche is a difficult car to steer at high speed. In my opinion, this does not hold true. It must be admitted, however, that the Porsche must be driven with a different technique than the normal front-engined car. This 'different' technique one has, so to speak, to learn. The Porsche is a car which announces in time when it reaches the limit of road adhesion and it is for this reason that I find it easier to drive than a normal car. There are many cars which possess good road-holding characteristics and which one can really drive to the limit of adhesion. Once this limit is crossed by just the slightest margin, these cars break loose and with such force that steering correction will hardly keep them on the road. The Porsche announces the fact that it intends to break loose, a little in advance, by a side wiping of the rear wheels and if one does the correct thing at this particular moment, then the first wiping away of the rear wheels becomes completely harmless. As a matter of fact, you will feel, from prolonged experience, that this is a completely normal and controllable action.

"Pleasure touring in a Porsche car will show little difference in handling characteristics from the normal automobile. However, in driving at really high speed there are two important points to remember. First is the motion on the steering wheel shortly before the turn and at the entrance of the turn itself; one must 'saw' the Porsche wheel. It is important to hold the steering wheel relatively loosely and make small corrective motions before the full centrifugal force has its effect on the car. One does not drive the car around the curve in a simple circular line—instead one drives it in a so-called 'snake line.' I would repeat that it is important to hold the wheel loosely, so that the 'sawing' motion does not become a jerking one, but remains a small continuous motion. You will find the steering gear of the Porsche helpfully easy and 'soft.'

"The second point is that when driving into a curve at high speed, one does not wait until the Porsche breaks away in a surprising thrust-like manner. On the contrary, intentionally and consciously one brings about the breaking away in the beginning of the curve, a few tenths of a second before the turn itself. This only has sense, of course, when the car is driven really fast, so fast indeed as to cause the car to break away regardless. What you really do is to force upon the car your intention in purposely making it break away, a command the car will follow obediently. After you have set up this situation, you don't have to recover the car as you would do normally by relatively strong counter-steering. The car moves as a whole, the front-end pointing towards the inside of the curve and it literally 'wipes' round the bend; this typical Porsche movement we call 'wischen' (wiping) and it is a state between normal rolling travel and skidding, and one makes the necessary correction by 'sawing' on the steering wheel with an easy hand.

"At the end of this wiping motion, one must watch that the car does not point too much towards the inside. In the ideal situation, the wiping motion runs out into a harmonious roll as one leaves the curve. This results in a high speed at the end of the curve because the whole wiping motion must naturally be a continuous acceleration. This is important because all braking must be completed at the beginning of the curve and you open the throttle earlier than with most cars, driving the entire curve under full acceleration.

"When one has understood this controlled wiping motion, better still, when its mastery has become part of your blood, then Porsche driving becomes enormous fun!"

The above description, by von Frankenberg, of the Porsche technique gives a very good explanation for the antics of a Porsche that is being driven fast. His remark that the steering is easy and soft is due to every steering box being "run-in" for 24 hours before being fitted to the car. The steering gear is conventional worm and sector running in engine oil, and in the assembly shop is a machine to take five assembled boxes and they are worked through the full range from lock-to-lock by a reversing mechanism driven by an electric motor. The lids are removed from the boxes and a special running-in oil is flowed through them. Day and night the steering box is wound from lock-to-lock and thus the full length of the worm is lapped to the sector, final adjustment being by shims on the sector pivot. This explains to a great extent the lightness of Porsche steering, its constant "feel" throughout its movement and its complete lack of high-spots or harshness, a feature that all who have driven a Porsche must have noticed. The "snake-line" referred to by Frankenberg is well summed up by a friend of mine who described the movement of a Porsche round a curve as a series of chords and tangents. The closing remark, about getting it into your blood, is very true, Porsche driving not only becomes enormous fun but you become a Porsche addict, like the addicts of chain-gang Nashes, Bugattis or Bentleys.

In almost exactly eight months of fun I have covered 30,000 miles with my "Damen" and executed every known Porsche antic with the exception of the "gound-level-flick-roll" and I find it hard to extract pleasure from other forms of motoring to the same degree. The performance is not the outstanding thing about the ordinary 1,500-c.c. normal touring Porsche; it will out-accelerate a Zephyr or Mark VII Jaguar, the maximum on English roads is a very honest 90 m.p.h., while given time, as on an autobahn or by-pass, it will do 95 m.p.h. and, under favourable conditions such as a mountain side or a following wind, I have done 103 m.p.h., these figures being rev.-counter readings, the speedo being 7 m.p.h. fast at maximum. The genuine top speed, such as I would do on the way home from the office every day, if I had to suffer such a journey, is 4,600 r.p.m. in top (95 m.p.h.) and with a maximum-permissible of 5,200 r.p.m. this is a comfortable feeling when a long downhill stretch approaches. These sort of figures are not the real charm of the Porsche. That lies in its manner of going, for the suspension gives a very smooth ride, the body makes negligible wind noise, the controls are light, the all-synchromesh gearbox is one of those that will go down in history, and a new standard in lightness, the engine emits a hum like a dynamo at a crusing speed of 4,200 r.p.m. in top, and the seats are comfortable, while the driving position always evokes cries of acclamation from anyone who tries it. It is only natural that a car developed around competition motoring should have an excellent driving position, and as the Porsche saloon is not easy to get into and out of, unless you are fairly agile, it is rather fun to watch an awkward friend struggling to get in and hear the muttered remarks about dwarfs change to satisfied comments as he finds a near-perfect driving position. Visibility, in spite of almost sitting on the ground, is such that even I can see the road eight feet in front of the bumper when adopting a fairly reclining arms-stretched position. If you sit upright, vintage-fashion, this is reduced to six feet. The easy, quiet manner of going and the comfort factor all combine to make it possible to cover more than 600 miles in a day's Continental motoring, driving alone, and yet not feel tired at the end, while 8-10 hours of a consistent average of 35 miles into each hour will show a fuel consumption of 34 m.p.g., the range of the tank being an easy 350 miles. On English roads 400 miles in a day are effortless and do not involve early starts or late arrivals.

Of all the "bogeys" thought up by anti-Porsche types the only serious one is always overlooked; most of the "cracks" at Porsches can be counteracted by the simple remark "When did you last drive a Porsche?" to which the answer is invariably "Well, I haven't *actually* driven one," so you follow up this with "When were you last taken for a ride in one?" and you would be surprised how many leaders of the opposition then have to admit that they've never been in a Porsche and they add feebly "but, I do know that it's a fact, old boy." Even quite well-known rally and racing drivers have been caught out by this counter-play and when they are the central figure of an admiring group they do not like it at all. If these "know-alls" had any experience of Porsches they would quietly ask about the health of the rear tyres and then it would be my turn to look embarrassed, for rear-tyre wear is a slight problem. My best pair lasted 6,500 miles and the worst 4,700 miles, and as I have used various makes there is little point in naming them, except that the best were some much-maligned racing tyres. This excessive wear is entirely dependent on how you drive; if you "wischen" all the time, as I do, then you must pay for it, if you are content to tour then you could do 20,000 miles. The VW is a similar sufferer, one owner needing new rear tyres in 8,500 miles, another having done 14,000 without a trace of wear. Equally, I knew a man who could never do more than 6,000 miles on his hotted-up Morris Minor without needing new tyres; speed round corners as well as high cruising speeds must be paid for no matter whether you over or understeer. If you can afford the fun, well why not have it. The Porsche front tyres are very reasonable, they last 12,000 miles.

As to maintenance and reliability I have few grumbles, for maintenance has consisted of routine oil changes every 1,500 miles, new Fram filter element every 10,000 miles and regular greasing. Reliability is such that I never stopped on the road for any "mechanical" reason in those 30,000 miles, though I did suffer some troubles. At 5,000 miles some of the grease in the speedo cable worked its way into the instrument and being of the magnetic-drive type it was converted to hydraulic drive, which meant that it indicated 120 m.p.h. most of the time, but as the mileometer was mechanical and continued to work I did not bother about the free replacement until I returned to the factory at 16,600 miles, the reasons for this return being manifold. I had got the hang of "wischen" motoring and had been warned that if I indulged in it on bad road surfaces the gear-lever would give a violent judder and the rear-end would be subject to abnormal strain and the gearbox mounting might crack. I did what I was told not to do all over the Italian mountains, through the Massif Centrale of France, the Pyrenees and the Portuguese mountains and on the return from Lisbon there was an ominous "click" each time I lifted my foot. This was after 16,000 miles, so I could hardly complain and, anyway, the only effect that this broken gearbox mounting had on the car was that it jumped out of top gear on very bad bumps at over 80 m.p.h. In addition I had a cracked windscreen collected from flying stones when overtaking another car, and I felt it was time the engine was looked at, while the rear shock-absorbers were worn out and I wanted new heavy-duty "competition ones" fitted.

While at the factory I agreed to let the Service Department give the car a routine 15,000 check, but when I suggested a decoke and valve grind they roared with laughter. I had to admit that it was still doing its 4,600 r.p.m. in top, used a pint of oil between changes, and ran as smoothly as when I left the factory, but being old-fashioned I felt the cylinder heads ought to be removed after 16,600 miles of very hard driving. All the mechanical units were removed from the car, cleaned, checked and replaced, though nothing was opened. The engine was checked for compression, timing, valve clearances (they had been done once, at 10,000 miles) and new plugs and ignition points fitted, but that was all. The brake linings were renewed and the drums checked for truth and skinned where necessary and new stub-axle assemblies were fitted. The brake linings and stub axle and suspension bushes were replaced as a matter of course, on an exchange plan, not because they were worn out but because the next check was not anticipated before 30,000 miles and by then the original ones might have worn a little and as the Porsche is a car meant to be driven fast, the Service Department like to know that everything is 100 per cent. While this work was being done I was able to see the new gearbox mounting, which is incorporated in the 1956 cars and will avoid the trouble I had, and also the new saloon with the Carrera engine. At the same time I had the opportunity to try a Speedster, the cheaper open two-seater, with normal 1,500-c.c. engine. Being more spartan it was considerably lighter, my car weighing 17½ cwt. in normal road trim, covered with odds and ends such as extra lamps, radio, tools and so on, and being really fully equipped for comfortable touring, so that this Speedster together with its slightly lower axle ratio was a very lively car. As most of the weight saving arose from it being open it resulted in a lower c.g. and 90 per cent. of this weight removal was from the rear-axle loading, being roof, windows and mechanism, rear window, rear-seat squab and lighter bucket seats and as a result there was quite a marked difference in handling, the "wischen" cornering not being so pronounced, but as it was a sporting two-seater it had all the failings of such models such as unlockable doors, flapping hood at 90 m.p.h., continual indecision about hood-up or hood-down, and while being great fun as a "dicer" it was not what I would have liked for motoring 1,000 miles a week continuously, that is, when the normal saloon Porsche is the available alternative. After all, one of the most pleasing things about the Type 356 saloon Porsche is its aerodynamic body, which spells efficiency and a sense of keeping abreast of the times; it is difficult to justify an open car for long-distance high-speed touring.

FUN MACHINE.—The Speedster can be had with Normal, Super or Carrera engine and as a sporting two-seater it is enormous fun to drive, being particularly lively in mountain country. The soft hood has a hinged frame and fits to the windscreen with over-centre clips, there being detachable perspex sidescreens to make it weatherproof.

While doing the routine Service Schedule, every little detail on the car was checked, even to fitting a new grille in the ash-tray; I had discarded the original as being a non-smoker I used the ash-tray for toffee-papers. Thrown in as a matter of courtesy was a check for alignment of the machine in the assembly department, tightening and a better fitting of the extra lamps I had mounted myself and a wash and polish, inside and out. The steering had required a 40 thou. skim to satisfy the meticulous Porsche standards and the car was given a 20-mile road test before being handed over for me to drive like a lunatic for another 15,000 miles over Europe's best and worst roads, conditions that would never prevail in England.

After a visit to Sweden I returned to Germany and competed in the Rhineland Rally, a wonderful event consisting of leaving the car in the open all night and from a 7 a.m. start covering 22 laps or 380 miles round the entire Nurburgring, using the Grand Prix circuit and the old Southern loop as well. The only stipulation was a maximum time for the whole distance and in the 1,600-c.c. Gran Turismo category the schedule was aimed to make a Porsche "Damen" hustle along a bit, though easy enough for a Super 1,500 c.c. Those who qualified then competed in a timed hill-climb on one part of the circuit. By now I had covered 21,000 miles and "race" preparation consisted of a new set of racing Dunlops, an oil change, and removal of all my luggage. After 17 laps around an average of 60+ m.p.h. I suffered a choked main jet, probably from the open refuelling churns used, and the time lost in locating the trouble after a slow return to the pits was more than I had in hand, so that was that. However, it was a good dice while it lasted.

Returning to duty motoring, the end-of-season trips brought the total to 28,000 miles and two weeks in England made up the round 30,000. Passing through Stuttgart on my way to England I paused to have new rubber bushes put in the front anti-roll bar as it was rattling when going over cobblestones and I again suggested a decoke, or new piston rings or something. The chief tester took the car out on the autobahn and could not see why I was complaining —well, I wasn't exactly complaining, I just thought . . . They took the clutch adjustment up a notch, clicked the brake adjusters a couple of notches each and I left thinking "I suppose they know best." On the way from Stuttgart to Cologne I put 210 miles in three hours, did 77 miles in the first hour, held 4,800 r.p.m. in top for at least three minutes on the Dormstadt-Heidelberg stretch that used to be used for record breaking, all at 33 m.p.g. with the radio playing, and realised that perhaps Dr. Ferry Porsche and his men do know something about building nice touring cars.—D. S. J.

A CARTER SPECIAL!

Sir,

As the owner of a Model-T Ford I was rather shaken to read in Mr. Carter's book, "Edwardian Cars," that the Model-T Ford drove on one rear wheel only and had no differential. The best bit, however, concerns the engine, quote . . . "In place of having cylinder tops cast with the block, a cover was provided which hinged along one side, so that by loosening a few bolts it could be *turned back*, thus exposing the pistons and valves." . . . see page 151. It would indeed be most interesting to know the present whereabouts of this amazing Ford!

I am, Yours, etc.,

London, E.18. HAROLD PRATLEY.

HOLT

We have been informed by Douglas Holt Ltd. that their product Silencer Seal costs 3s. 6d. a tin, and not 2s. 6d. as quoted in their advertisement in last month's issue.

South German Test-outing

CRESTING a rise on the Autobahn spur which leads to Heilbronn, the Porsche Spyder competition 2-seater for which over 135 m.p.h. is claimed has headlamps blazing to warn other traffic of its approach.

A TRIO OF PORSCHES SAMPLED

By Joseph Lowrey, B.Sc.(Eng.)

BETWEEN the B.M.W. and Porsche companies there is at least one thing in common, the fact that both started virtually from scratch after the war. Whereas the B.M.W. factory at Munich is setting out to serve a really big market with the Isetta coupé, however, the Porsche factory in a Stuttgart suburb is dedicated to the much more specialist task of producing sports cars.

From its inception in modest premises where Volkswagen components were built, into a sports coupé of finely streamlined shape, however, the Porsche company has advanced a very long way. In the cars, ideas inherited from the late Ferdinand Porsche's design for the Volkswagen are still evident, but of actual VW components there are now virtually none. Likewise, the original factory has now given place to a new one close at hand which, while still moderate in size, makes very bold use of a hillside site to employ eye-catching yet practical unorthodoxies of architecture. Porsche sell more than half their production to Americans, many to those stationed in Germany, who can therefore visit the factory.

Work as consulting engineers on various projects is still continued actively by the Porsche company, a prototype military vehicle with air-cooled rear engine and four-wheel drive having already been unveiled, and an air-cooled V-type engine being evident in one corner of a works normally associated with flat-four power units. Sports cars are the main business, however, graduated from normal road-going coupés and cabriolets through those similar-looking cars with two more potent types of engine to the "Spyder" 2-seater which is designed primarily for racing.

Production capacity in the Porsche works is quoted as 14 cars per day, although at the time of our visit the recently announced 1956 models were as yet only emerging at about half this rate, due to delays in obtaining castings and other outside components. Quite elaborate tooling for the machining of cylinder blocks, crankcases and other parts was evident, but the policy of having one craftsman responsible for assembly of a particular engine almost from scratch is adhered to, the neat and clean benches being most impressive. Trimmed bodies (this is a chassisless steel car) from the Reutter factory next door are mounted on high trolleys, and moved in line through the works as mechanical components are assembled on to their underside, each car then undergoing about 60 miles of testing on the nearby Autobahn before delivery. "Trade plates" mounted on an apron which covers the whole front of the car guard against damage to paintwork during these tests.

Most buyers of Porsche cars are urged strongly by the makers to choose the normal engine, which uses plain big-end bearings. The rather similar looking Super engine with roller big-end bearings and higher power output offers more performance for use in competitions, but has

NORMAL version of the latest 1.6-litre coupé, the Porsche seen here proved itself a 100 m.p.h. car in which the low level of wind noise was especially notable.

a shorter life between overhauls and is noisier and less flexible, disadvantages which are thought to outweigh extra speed for most buyers. The Carrera model, with a slightly modified version of the "Spyder" four-camshaft engine is even more specialized.

Externally the Porsche changes comparatively little from year to year, care being taken to avoid spoiling a good basic shape. Internally, it has altered a good deal for 1956, perhaps the most valuable change being an enlargement of the cylinder bore to step up the displacement from 1,488 c.c. to 1,582 c.c. Also, the torsion bar suspension has been softened quite considerably and provision made for longer wheel travel, softer springing increasing the effectiveness of the front-end anti-roll torsion bar which was introduced last year, and being matched by improved shock absorbers.

With a revised model not yet in full production, up-to-date cars were scarce around the factory, but after a drive in the 1.5-litre car which Klaus Rucker (a Porsche engineer lately returned to Germany after a period with Studebaker in America) was using for personal transport, we were chased away by von Hanstein for a spell of motoring in a 1.6-litre car, which had not yet finished routine testing before the carpets were installed and it was delivered to a dealer.

Having quite a personal liking for cars which can be "wished" around corners without appreciable physical "steering," my first reaction was to prefer the handling of the 1955 car to the slightly more "Americanized" controls of the 1956 model. After a wild dice through the back-streets of Stuttgart in the latter, however, with von Hanstein in the passenger seat urging me to greater efforts in the hope that we might get ahead of Uhlenhaut who was hustling more conventionally down the busier main road in his Mercedes 300 SLR coupé (we failed by a narrow margin) I soon forgot my reactionary ideas.

By the standards expected in England, the Porsche is by no means flexible in its performance. On the rev. counter there is marked a green sector from 2,500 r.p.m. to 4,500 r.p.m., a range of "desired" speeds which means from 13 to 24 m.p.h. in 1st gear, from 24 to 43 m.p.h. in 2nd gear, from 37 m.p.h. to 67 m.p.h. in 3rd gear, and from 51 to 93 m.p.h. in top gear. There is not much overlap in these speed ranges, to give a driver freedom of

FASTEST product of the Porsche factory, the Spyder 2-seater has its engine ahead of the rear axle instead of behind it, and uses a multiple-tube chassis frame whereas touring models have combined steel bodies and chassis.

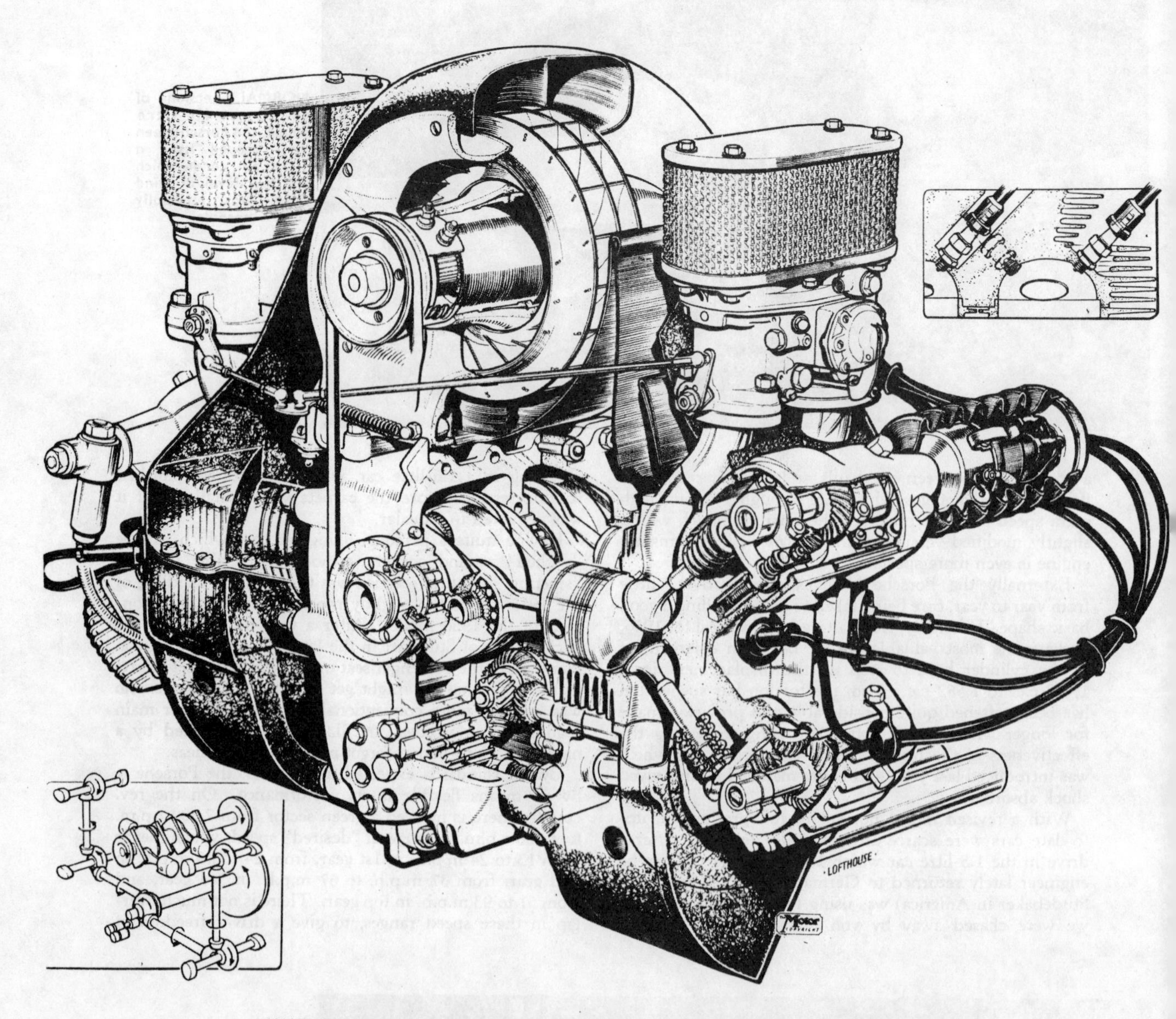

AVAILABLE with small modifications in normally bodied cars, the Spyder competition engine is shown in this drawing by an artist of *The Motor*. Air cooling of four horizontally opposed cylinders is retained, but pushrod valve operation gives place to four overhead camshafts driven by shafts and bevel gears. Roller main and big-end bearings are used on the Hirth built-up crankshaft, carburation is by a pair of twin-choke downdraught carburetters, and two plugs per cylinder are sparked from distributors on the rear ends of the two inlet camshafts.

choice about when to change gear upwards or downwards. In actual fact, however, with the larger and more flexible 1956 engine the speed can drop as low as 1,800 r.p.m. before snatch becomes evident, and the red sector on the rev. counter from 4,500 r.p.m. to 5,000 r.p.m. seems to be intended merely as a suggestion that sustained cruising at these speeds is not advisable.

My impression, incidentally, was that it was not really the engine which was in any way inflexible, but that the use of a short and "solid" transmission line was probably making the use of low r.p.m. a jerky process. A spring-centre clutch, long propeller shaft and axle mounted on "cart" springs together provide flexibility which if cleverly used can reduce the "resonant" speed of a transmission to a very low figure.

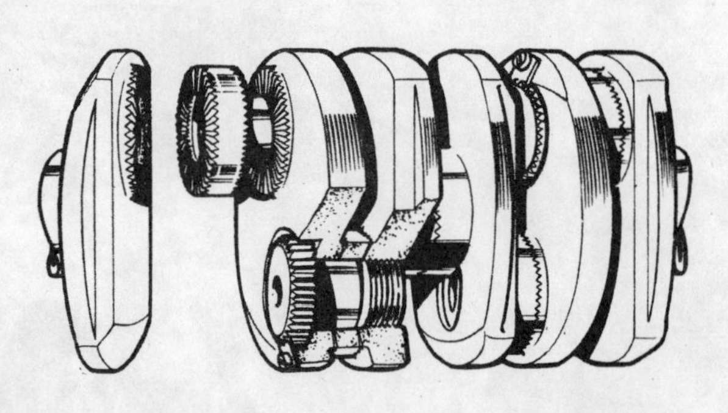

STARK simplicity characterized the cockpit of the sports-racing Spyder which was sampled, although some cars are built with touring equipment for use in Rallies. Location of the engine immediately behind the driver, with the spare wheel enclosed above the gearbox, will be noted.

using an indicated 5,000 r.p.m. in each gear, the timed acceleration figures were: 0-30 m.p.h. in 5.1 sec., 0-40 m.p.h. in 7.7 sec., 0-50 m.p.h. in 11.1 sec., 0-60 m.p.h. in 16.6 sec., and 0-70 m.p.h. in 25.3 sec., with a standing quarter mile covered in 19 sec. Balancing the newness of the car against its having slightly less weight aboard than it would carry if subjected to a proper Road Test Report, these are commendably good figures, especially the standing ¼ mile time which benefits more than the others from a no-wheel-spin getaway with plenty of weight on independently sprung rear wheels.

Whilst the Porsche is not exactly silent mechanically, it is most notably free from wind noise when running quickly on the Autobahn, and the new springing gives a comfortable ride without introducing perceptible roll to the cornering characteristics. There *is* a back seat, and to record acceleration times I sat in (or more precisely, across) it, after the driving seat had been slid well forward, but I was very glad to escape from that cramped place at the first opportunity. Call it a two-seater, with room at the back for small children or a lot of luggage, nicely furnished as a businessman's rushabout.

Even Faster

One more outing was provided for us before we left Stuttgart, an outing which contrasted strongly with the comfortably brisk travel in the Porsche 356A/1,600 coupé. One of the four-camshaft Spyder racing 2-seaters, with its engine mounted within the wheelbase and carrying very simple full-width bodywork on its tubular chassis, was held back for a spell from being stripped for a change of axle ratio, and we were each given a ride up and down the Autobahn. Traffic did not allow the speedometer needle to go round to the claimed top speed of 220 k.p.h. (137 m.p.h., on 1½ litres), and without the shelter of even an aero screen in front of the passenger seat the tasks of keeping on breathing and of avoiding being blown into a reclining position distracted us from any attempts at very scientific observation. But, if the speedometer was to be believed, the great surge of acceleration from rest had not quite faded out as 200 k.p.h. (124 m.p.h.) was exceeded, the works test driver complaining that the bulk of a passenger was slowing the car considerably. With this same engine, slightly decompressioned and de-tuned but quoted as giving only 10% less power, installed in one of the sleek hard-top cars, remarkably impressive performance should be provided, although traffic might be quite something of a headache.

Unexpectedly, we found that our return journey coincided with a one-day public holiday in Germany, which frustrated a few odds and ends of shopping but provided compensations by greatly reducing the number of heavy lorries on the Autobahn. After a detour on to ordinary roads across the fog-filled Rhine valley to Bonn for lunch, and a teatime stop at Arnhem, we reached the Hook of Holland in comfortable time to eat a very adequate meal in the station restaurant, then embarked early for the return to an England which, while we had been "enjoying" alternate fog and drizzle, had unkindly staged in our absence some of the kindest weather of the winter!

What makes a relatively narrow band of really useful engine speeds perfectly acceptable is, of course, the fact that there is no undue fuss at the higher r.p.m., and that the gearbox is a delight to use. The brand-new engine of the car we were driving would not mind short periods of hard driving, we were assured, so we decided to take a few performance figures, two-up but without our usual weighty box of test apparatus, along the least-hilly piece of the Stuttgart to Heilbronn Autobahn which we could find. Checking the kilometres-per-hour speedometer at speeds between 30 and 80 m.p.h. equivalent, we found fairly consistent optimism of 6% or 7%, so having found an indicated 170 k.p.h. to be attainable either way on the level decided that this (non-Super) model was indeed just about a true 100 m.p.h. car as claimed.

Against the Watch

So far as acceleration went, the 1,600 c.c. Porsche had very nearly constant urge through its modest range of useful speeds. Space between road undulations was limited, but the two-way average time from 30 to 50 m.p.h. was 15.4 sec. in top gear or 8.5 sec. in 3rd gear, the corresponding times from 40 to 60 m.p.h. were 16.6 sec. and 9.1 sec. respectively. From a standing start,

Even in this five-minute pose beside Stuttgart airport, the Porsche looks impatient to get on the move

The Autocar
ROAD TESTS

No. 1594

PORSCHE 1600

ON March 16 the firm of Dr. Ing. h.c.F. Porsche celebrated in Stuggart its 25th anniversary. Most of those years have been spent helping other manufacturers on design problems, and in the manufacture of diesel engines which continues. Until recently the late founder was best known as the man behind the pre-war Auto Union racing cars, and as the designer of the Volkswagen. Then came the introduction of the Porsche sports car in 1949, based extensively on Volkswagen parts. Steadily the car has developed into a complete Porsche entity and now, with the founder's son at the head, the factory produces a range of models that are internationally respected.

Fastest in the family is the 1½-litre, 110 b.h.p. sports racing Spyder, an open two-seater; then come the Carrera-engined coupé, convertible and speedster, powered by the twin o.h.c. engine developing 100 b.h.p., also from 1½-litres, and the same body styles with the 1,600 c.c. o.h.v. unit. There are also the coupé and convertible with a 1,290 c.c. engine. The 1,600 in standard tune develops 60 b.h.p., while the Super develops 75 b.h.p. at the expense of being relatively a little rough. The Carrera-engined cars are very expensive, and made only to special order.

A test of the 1,500 c.c. coupé appeared in *The Autocar* of November 6, 1953, and as it is the standard 1,600 that now constitutes the bulk of production, this model was selected for a full test, recently completed in Germany.

Apart from the small increase in engine size and the effect produced thereby, the 1,600 is very similar to the model previously tested. That is, both have rear-mounted, air-cooled, horizontally opposed o.h.v. engines, and smooth body shapes designed for two adults and one or two children (or a third adult on local journeys).

The sports-racing background to the Porsche is discernible from the moment the car moves off. The driver realizes at once that this model is something right out of the run of ordinary cars. The placing of the controls, the seating position, and the acceleration provided by the good power-to-weight ratio, are but a part of the first impressions. To these can be added the high gearing, superb gear-change mechanism, and steering accuracy of the highest order. Here is character that is not easy to define, conveyed indirectly perhaps, and in part by the international custom of Porsche drivers to flash head lights at each other in salute.

Porsche claim that any 1,600 will achieve 100 m.p.h. on the flat, and comment that most cars, after owners have completed the running-in process that is well advanced in the factory, will achieve higher speeds. The car tested achieved its 100-plus, compared with a speed of 91 achieved with the 1,500, but the greater smoothness of the engine is also an important difference between this car and the earlier models. Also the handling has less "rear engine feel" to it.

In its present form the Porsche behaves more like an orthodox high-performance sports car, although a certain skittishness at the rear, attributable partly to the swing axle rear suspension, can still be felt. In this connection it must be made clear that the stability remains very good indeed, and the design as a whole gives a liveliness to the controls of which the skilled driver can take advantage.

In Germany it was possible to enjoy the car on *autobahnen* and ordinary roads. Currently the German *autobahnen* carry heavy traffic, and the behaviour of the lorry drivers is unpredictable. Thus the Porsche is more than fast enough to keep its driver alert on such great highways. Even an *autobahn* has cobbled hills, with curves sharper than usual in the mountainous areas, and with the rev counter indicating 4,500 r.p.m. or more, and the speedometer comfortably above the kilometre equivalent of 90 m.p.h., these cobbled downhill curves can be negotiated with an absolute stability that flatters any driver who has a light touch.

Off these main highways the car is even more enjoyable, for its modestly sized engine does so much without fuss, provided that the driver expends on the gear box energy equivalent to a housewife cutting soft butter. With 48

Speedy purpose is suggested by the frontal appearance. Side lights which incorporate flashing indicators are mounted beneath the head lamps, and there are outlets connecting with the horns. There are no extraneous protrusions to cause wind noise or reduce maximum speed

m.p.h. available on second without exceeding the red zone on the rev counter (which finishes at 5,000), and 75 available on third, top remains unused on winding or hilly sections, yet there is still no fussiness from engine or gear box.

Third is so quiet that even on the *autobahnen* it is possible to cruise behind other traffic at 70, thinking that the car is in top. Up and down hill, round every type of corner, seizing brief opportunities to overtake safely, the gear box is steadily in demand, and for this reason it is worth taking a closer look at it. All four speeds have synchromesh, and as the maximum speed on first is 29 m.p.h., this gear can be used when required on steep, hairpin turns, or when picking up after slowing to a crawl. The slim change lever is mounted centrally, but the box is at the rear. At a standstill, when engaging first or reverse, the distance between the lever and the box can be noticed by the behaviour of the mechanism, but once under way, changes are limited in speed only by the driver's agility, and the synchromesh is quite unbeatable. In second gear, and on the over-run, increased noise is noticeable.

Many high performance cars are often seen to be driven quite slowly, but observation of many Porsches in their native land reveals that these cars are bought by people who like to drive fast. Thus engine noise at high speed, and wind noise, too, are of importance. Regardless of road speed the 1,600 engine is never obtrusive, the noise level certainly being restrained for a car with such a high performance for its engine size. The really happy engine cruising speed that can be kept up for hours (on *autobahnen* at quiet times) is 4,000 r.p.m. (about 80 m.p.h.). Above this speed the engine note becomes more insistent, but many miles can be covered at full throttle without the oil temperature rising too high. The beautifully smooth shape of the coupé coachwork cuts wind noise down to the negligible category, and if the windows are closed conversation can be enjoyed in normal tones at maximum speed. When a window is open, discomfort is caused at high speed by reverberation of the air in the car which hurts the ears; because it is desirable to keep the windows closed, some improvement in ventilation in warm weather is desirable. As the car will so willingly cruise at 90 m.p.h. it is made to do so by most drivers when traffic conditions permit, making the m.p.g. range of 29-36 all the more creditable.

At under 1,500 r.p.m. the characteristics of the flat four engine and the transmission result in a reduction in response and a slight feeling of snatching. In traffic, free use of the gears is thus necessary to obtain flexibility of performance.

Oil temperature is important in the Porsche. If maximum speed is maintained for many miles, the temperature rises to about 120 deg C, which most owners would regard as the top permissible limit. One must ease up on super highways when traffic is light, but rarely are traffic conditions such that top speed must be limited by the rising oil temperature. When driven hard at high r.p.m. the oil is "thin," and owing additionally, no doubt, to fairly generous lubrication of the cylinder walls, the oil consumption is quite heavy.

Of all components the brakes play a major, if contradictory, role in giving speed and Porsche a synonymity. On test, maximum braking power was obtained at all speeds with no more than 50 lb pressure on the brake pedal—pressure, in other words, that is firm but not hard. The percentage efficiency was high, and the car pulled up all square even after repeated brake application. There was no trace of fade regardless of the type of driving. It is a comfortable feeling in this type of car to know that the brakes are always adequate, provided only that the road surface is not slippery, and in the 1,600 one has that confidence in full measure.

The model is very much a quality car, assembly of the mechanical components being completed with as much attention to detail as is the construction of the Reutter coachwork. All steering mechanisms, for example, are run-in on the bench from full lock to full lock for the equivalent of 5,000 km on the road. With only two and a half turns from lock to lock, and a perfectly smooth mechanism completely free from play, the steering is ideal, high-geared precision not affecting the light feel. Road shocks pass almost unnoticed, yet the feel of the steering is essentially live.

Considerable attention is paid to detail in the general conception of the coachwork, the most important aspect being the driver's seating position. Both of the separate front

The smooth shape is seen to best advantage from the side. There are no unnecessary curves, and ornamentation is restricted to a practical rubbing strip at the base of the body. Here the car is seen outside the Porsche administration block in Germany. Impressions on British roads of a 1600, provided by the British concessionaires, A.F.N., Ltd., of Isleworth, Middlesex, appeared in our issue of Nov. 18, 1955

The air-cooled engine is installed compactly at the rear. The fan looks after cooling and can also direct hot air to the interior of the car. Carburettors, dipstick and oil filler are all reached easily, and the engine can be removed for major service with little difficulty

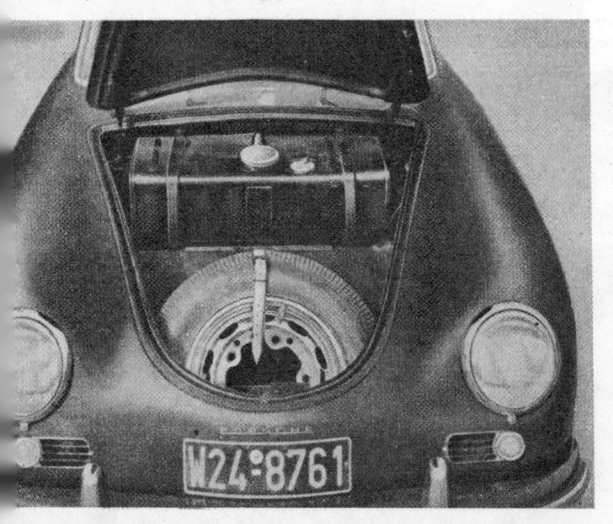

Luggage space is modest and of an awkward shape. Fitted cases would be necessary to take full advantage of the room available. The tools are housed at the foot of the spare wheel

PORSCHE 1600 . . .

seats have backrests adjustable from the vertical to semi-reclining positions, and the comfort is a happy compromise between the firmness of the sports-racing seat and the softness appropriate to the tourer. The passenger is not quite so well accommodated, as he instinctively swivels his shoulders away from the driver to provide maximum elbow room, whereupon his left knee (on a left-hand drive car) tends to interfere with quick engagement of top.

All-round visibility is good. The sharply sloped bonnet reveals the road to within a few feet of the driver himself and the car can be placed accurately without conscious effort. The windscreen pillars are slim and the area of glass in the sides is large. A considerable quantity of luggage can be carried within the body when the car is used as a two-seater without affecting rearward vision.

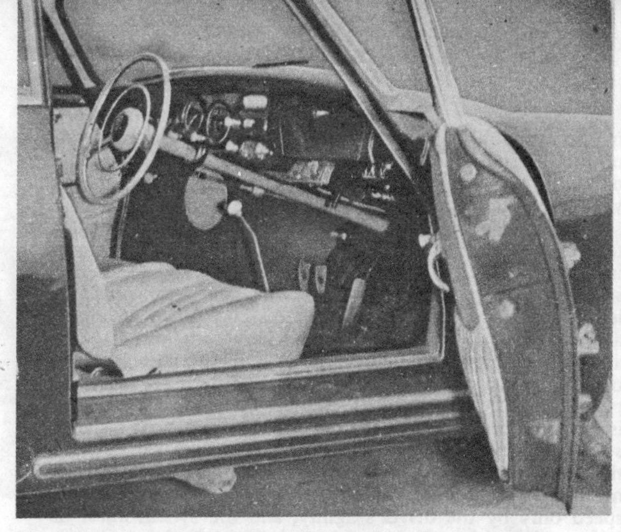

The layout of the instruments and controls is good, and the driving position excellent. The slim gear lever operates a gear box with unbeatable synchromesh on all four forward gears. There is a large radio speaker on each side of the scuttle. One must stoop to get into the car, but the wide doors make entry reasonably easy

A neat chrome grille allows air to escape from the engine compartment. The flashing indicator lights are separate from the orthodox rear lights, and there are deep overriders on the wrap-round bumper. Models with the standard 1,600 c.c. engine may be identified by the single exhaust pipe

Instruments and controls are well arranged. Directly in front of the driver are the important rev counter and the speedometer (with trip mileage recorder), and at the top of a third dial is the oil thermometer with the fuel level gauge below it. There are indicator lights for the ignition winkers, oil pressure and main head lamp beams. A reserve fuel switch is under the facia and foot operation is used for the windscreen washers.

The central button on the steering column flashes the head lamps in daylight and, when the lamps are in the dipped position at night, they bring on the main filaments additionally. A horn ring is fitted for audible signals. This button for lamp flashing is widely adopted by Continental manufacturers and is most useful at any hour. (It also explains why Porsche drivers can effect an understanding flash at each other at very short notice.) The standard head lamps are powerful, enabling high speed to be maintained during the night. Although they are mounted lower than on most cars, the range in the dipped position is adequate and does not seem to worry oncoming drivers.

Detail finish of the Reutter coachwork is of a standard associated only with high-quality cars and all fittings are solidly made, work smoothly, and are well finished. Yet extensive use of light alloy helps to keep the kerb weight down to 16 cwt.

There is a lockable glove compartment in the facia and slim pockets in the doors and below the radio speakers on either side of the scuttle. An ashtray is fitted centrally in the facia above the radio controls and above it is a good map light which, however, is not screened from the driver's

There is little room in the rear for passengers, partly because the backrest is too near the vertical. However, an adult can be squeezed in on short journeys, or children can be accommodated. The backrest of the rear seat folds down to allow extra luggage to be carried

PORSCHE 1600 . . .

eyes. The trim and upholstery are of better quality than those usually found on Continental cars.

Some luggage can be accommodated between the spare wheel and fuel tank under the bonnet. The space is of a rather awkward shape, for which fitted bags would be an advantage. When only two people are carried the backrest of the occasional rear seats can be folded down to make a platform, substantially increasing the total luggage space. The discomfort of an adult squeezed into the rear compartment for short journeys could be eased if the backrests were sloped more rearwards as, sitting sideways, a passenger suffers from the proximity of the front and rear seatbacks.

Engine accessibility is particularly good. The dip stick and oil filler are easily reached and there is no difficulty in attending to the carburettors and electrical components. For major service the engine must be one of the easiest to remove.

Partly because an engine of modest size that provides three-figure speeds and outstanding acceleration always wins affection, the superbly controllable Porsche brings back to motoring some of the joy that those privileged to drive sports cars in the earlier spacious days must have experienced. At the wheel one feels to be one up on the other fellow in all the things that matter in driving for its own sake. The imposition of duty and purchase tax make the total price formidable for British buyers, but the car remains, none the less, highly desirable.

PORSCHE 1600

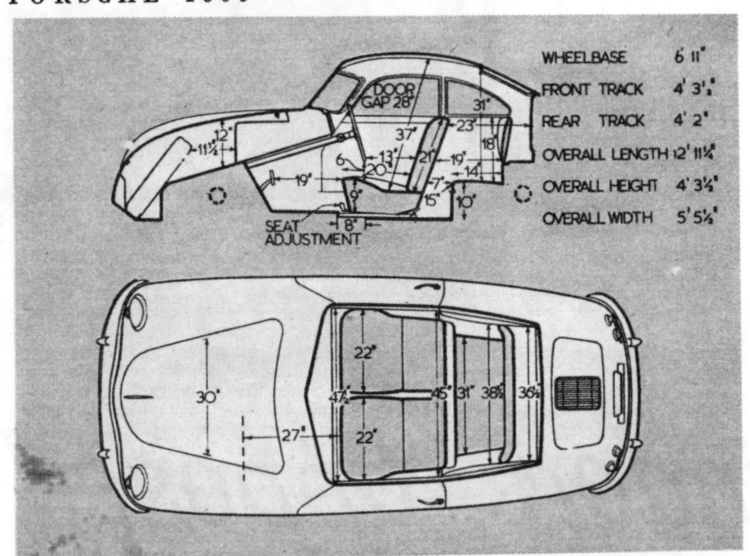

Measurements in these ¼in to 1ft scale body diagrams are taken with the driving seat in the central position of fore and aft adjustment and with the seat cushions uncompressed

PERFORMANCE

ACCELERATION: from constant speeds.
Speed Range, Gear Ratios and Time in sec.

M.P.H.	4.43 to 1	4.6 to 1	7.13 to 1	11.24 to 1
10—30	—	—	5.2	3.3
20—40	13.8	8.7	4.9	—
30—50	14.0	8.5	5.9	—
40—60	13.8	9.2	—	—
50—70	15.6	10.6	—	—
60—80	17.2	—	—	—

From rest through gears to:

M.P.H.	sec.
30	4.9
50	10.6
60	15.3
70	20.8
80	30.4

Standing quarter mile, 19.5 sec.

SPEEDS ON GEARS:

Gear		M.P.H. (normal and max.)	K.P.H. (normal and max.)
Top	(mean)	101.25	162.5
	(best)	102	164.1
3rd		59—75	94.9—120.7
2nd		38—48	61—77
1st		21—29	33.8—46.7

TRACTIVE RESISTANCE: 30 lb per ton at 10 M.P.H.

TRACTIVE EFFORT:

	Pull (lb per ton)	Equivalent Gradient
Top	170	1 in 13.1
Third	275	1 in 8
Second	440	1 in 5

BRAKES:

Efficiency	Pedal Pressure (lb)
63 per cent	25
85 per cent	50

FUEL CONSUMPTION:
31 m.p.g. overall for 765 miles (9.21 litres per 100 km.).
Approximate normal range 29-36 m.p.g. (9.7-7.9 litres per 100 km.).
Fuel, First grade.

WEATHER: Dry, sunny, slight breeze.
Air temperature 55 deg. F.
Acceleration figures are the means of several runs in opposite directions.
Tractive effort and resistance obtained by Tapley meter.
Model described in *The Autocar* of November 18, 1955.

SPEEDOMETER CORRECTION: M.P.H.

Car speedometer	10	20	30	40	50	60	70	80	90	100
True speed	12	20½	29	40	47	58	70	80	90	99½

DATA

PRICE (basic), with fixed-head coupé, £1,260.
British purchase tax, £631 7s.
Total (in Great Britain), £1,891 7s.
Extras: Radio £38, plus £17 3s 2d purchase tax.

ENGINE: Capacity: 1,582 c.c. (96.5 cu in).
Number of cylinders: 4.
Bore and stroke: 82.5 × 74 mm (3.25 × 2.91in).
Valve gear: o.h.v.
Compression ratio: 7.5 to 1.
B.H.P.: 70 at 4,500 r.p.m. (B.H.P. per ton laden 73.7).
Torque: 81.2 lb ft at 2,800 r.p.m.
M.P.H. per 1,000 r.p.m. on top gear, 20.

WEIGHT (with 5 gals fuel), 16 cwt (1,819 lb).
Weight distribution (per cent): F, 44.6; R, 55.4.
Laden as tested: 19 cwt (2,128 lb).
Lb per c.c. (laden): 1.3.

BRAKES: Type: F, Two-leading shoe. R, Leading and trailing.
Method of operation: Hydraulic.
Drum dimensions: F, 11in diameter; 1.57in wide. R, 11in diameter; 1.57in wide.
Lining area: F, 61 sq in; R, 61 sq in (128.4 sq in per ton laden).

TYRES: 5.60–15in.
Pressures (lb per sq in): F, 18.5; R, 23.0 (normal). F, 21.5; R, 25.5 (for fast driving).

TANK CAPACITY: 11.5 Imperial gallons (plus 1 gallon reserve).
Oil sump, 8.8 pints.
Cooling system, air-cooled by fan.

TURNING CIRCLE: 36ft. (L and R).
Steering wheel turns (lock to lock): 2½.

DIMENSIONS: Wheelbase: 6ft 11in.
Track: F, 4ft 3.4in; R, 4ft 2.1in.
Length (overall): 12ft 11.3in.
Height: 4ft 3.5in.
Width: 5ft 5.6in.
Ground clearance: 6.3in.

ELECTRICAL SYSTEM: 6-volt; 84 ampère-hour battery.
Head lights: Double dip; 35–35 watt bulbs.

SUSPENSION: Front, independent, swinging arms, anti-roll bar. Rear, independent with swing axles.

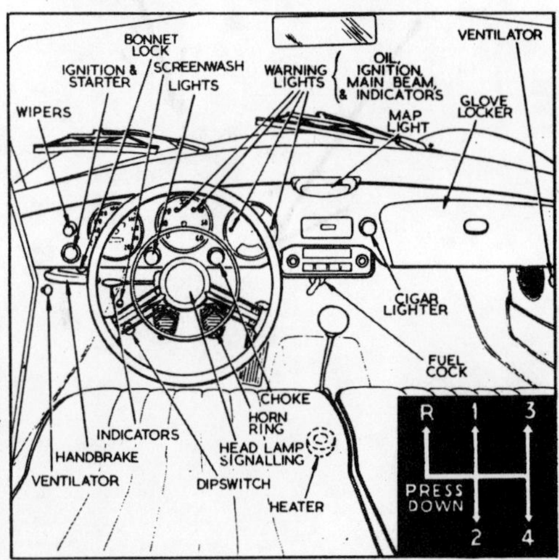

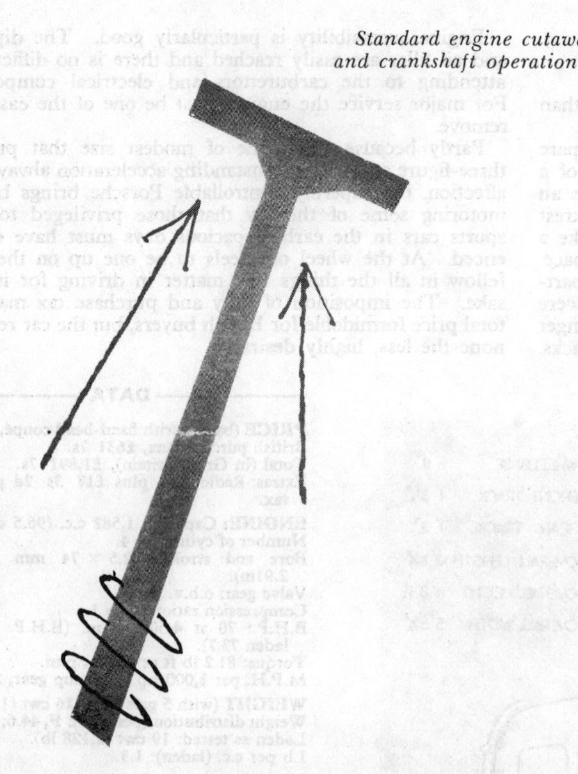

Standard engine cutaway shows cam and crankshaft operation

Four Cylinders Opposed

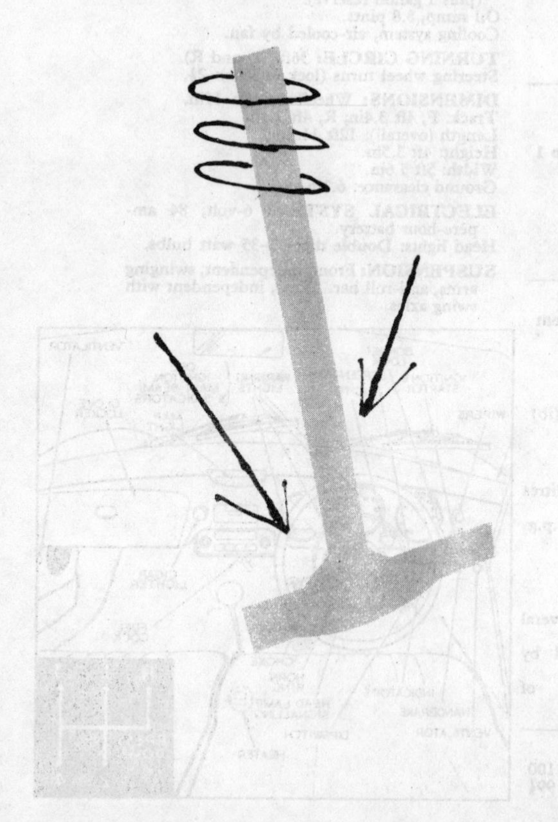

A sort of ultimate simplification, Porsche's flat fours are among the hottest in the world.

THE large, varied line of Porsche engines represents a series of evolutionary refinements based on the humble VW power plant which also, of course, is an original Porsche design. When the full range of variations on this unorthodox theme is viewed from its People's Car beginning to its present competition-car culmination, we can begin to appreciate the really earth-shaking significance of the Porsche *boxer motor.* It's an amazing device from the standpoints of originality, versatility, simplicity, durability, efficiency, and performance. Its many forms are naturally confusing to anyone who has not been able to

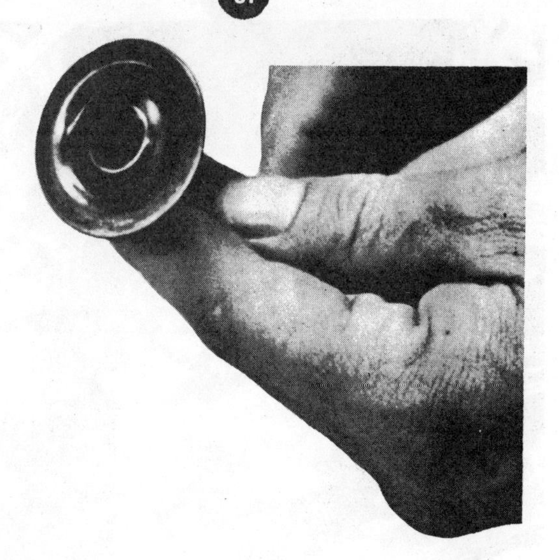

Exhaust valve has high center so that if piston strikes valve, it will close it, not bend stem.

By GRIFF BORGESON

Cylinder, piston and rod assembly from 1600 engine. This is not a roller bearing connecting rod. Note oil ring below wrist pin.

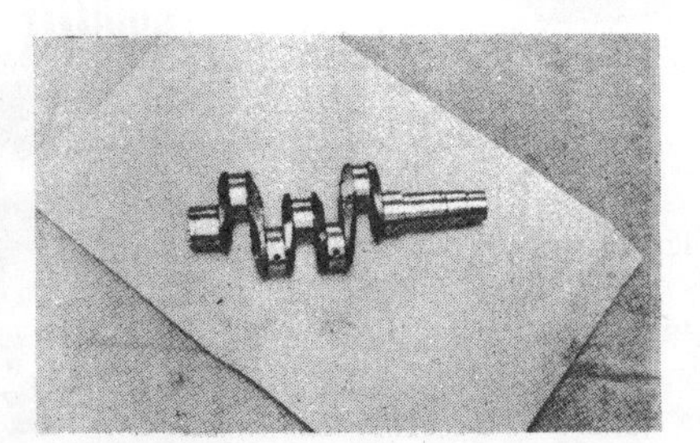

Standard crankshaft journals are narrow but, massive, with the mains running just under two inches. Bearings are hardened.

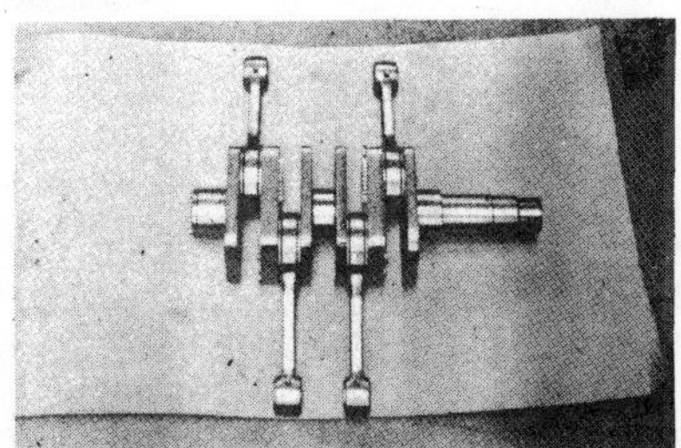

Hirth built-up roller-bearing crankshaft. Con rods are one piece and hand finished. Replacement price of unit—$513.

Spyder 1500 RS engine being prepared for Sebring. The dohc engine uses dual-throat Weber carburetors, and two separate ignition systems. Note small flywheel.

make a study of the subject and so we present here, for the first time, a survey of the whole spectrum of Porsche power plants.

The current line of engines manufactured by the firm called Dr. Ing. h. c. F. Porsche KG consists of the types shown in the accompanying table, the popular 1500 and 1500S having been replaced by the more lustily-endowed 1600 series. The 1100, 1300, and 1300S are paired with the coupe and convertible bodies. In addition to this choice of coachwork, the 1600 and 1600S also are supplied in the Speedster, a roadster or open two-seater. The same applies to the 1500GS Carrera. The appallingly potent 1500RS, with its unique ahead-of-rear-axle mounting, lives only in the Spyder, an all-out competition two-seater.

All of these engines share many of the same components. Some of these are crankcase, cylinder barrels, and cooling system. All of the "normal" engines — the 1100, 1300 and 1600 — share the same crankshaft, connecting rods, camshaft, and valve train. The S or Super engines all have Hirth roller-bearing crankshafts and special, ground-all-over connecting rods. The 1500GS and RS also have these organs, plus dual overhead camshafts, twin-throat carburetors, two-

Partial cutaway of the 1600S engine shows intake and exhaust ports. Intake valve is .28 inches larger than the exhaust. Piston has flat top, bevelled edges, and exhaust valve relief.

View of Spyder engine from fan pulley end. Timing degree marks can be seen on pulley. Engine is timed statically with 24° advance for each bank of cylinders. Advance marks must be at top center and aligned with vertical casing line.

plugs-per-cylinder heads, and a separate ignition distributor for each set of plugs.

A moment's contemplation of the table shows the fantastic range over which this engine of such modest beginnings has been made to perform and, as the world knows, to perform well. Every forward step in the performance of Porsche cars has caused endless bother and embarrassment to the people who build the engine in its most economical form. From the earliest days of the Porsche the VW factory was assailed with customer complaints, all hewing to this line: "If Porsche can get more speed and acceleration from the VW engine, why can't the VW factory do the same?" The answer, of course, lies in the difference in price between VW and Porsche cars.

In the Porsche 1600 road test I mentioned that calculations based on pulling power and tractive resistance indicated that the test car was delivering more horsepower to the flywheel than its builders claimed for it — a very unusual occurrence. This caused me to doubt my own slip-

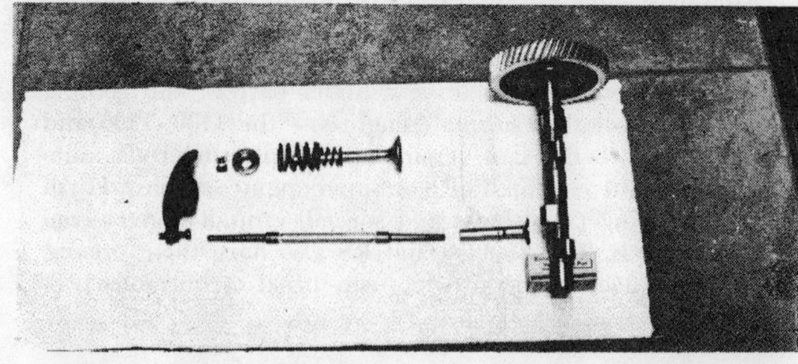

Valve train of standard engine. Pushrod is of light alloy with steel center section. Rods have same coefficient of expansion as cylinder-head assembly. Note blunt ends of camshaft. Valve clearances are .006 int. and .008 exh.

Bottom view of standard cylinder head which carries steel or bronze valve seats, valve guides, and plug bosses.

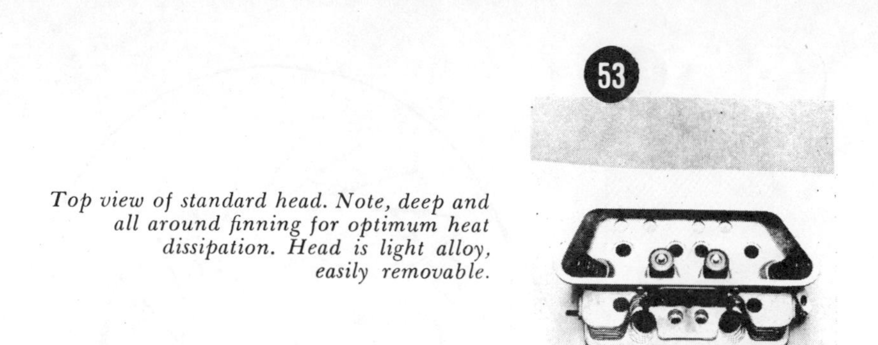

Top view of standard head. Note, deep and all around finning for optimum heat dissipation. Head is light alloy, easily removable.

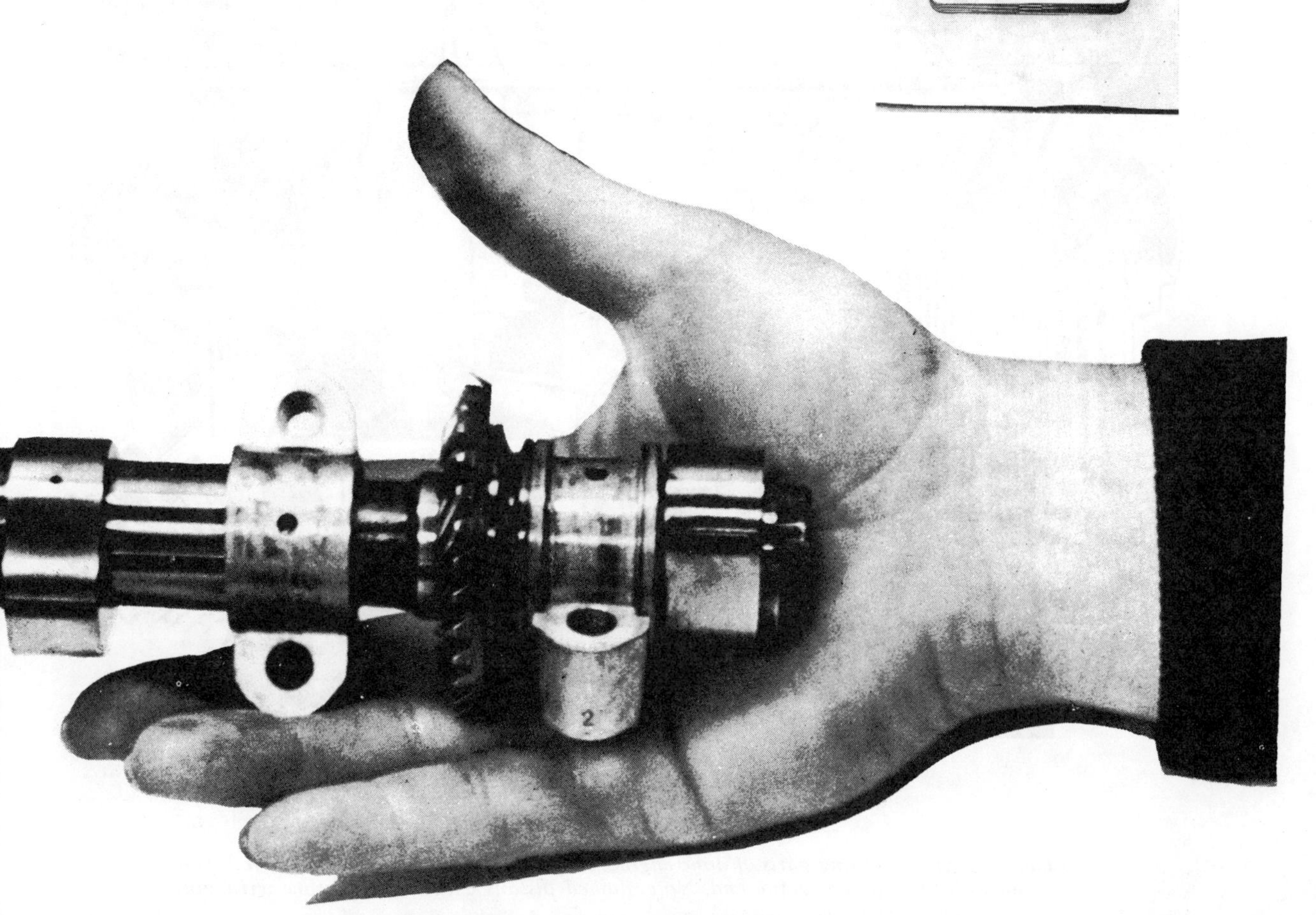

Part of overhead camshaft showing cam lobe, bearing, bevel gear, bearing, and second cam lobe. This cam operates the exhaust valves.

stickery and make inquiries. I learned from several owners who have had their cars on chassis dynamometers that they have run into the same surprising situation: power output was "a lot more than it ought to be." Finally I met Mr. Rolf Wuetherich, master mechanic from the Porsche factory who looks after company affairs in the American southwest. "Sure," he said. "It's our policy to keep all statement concerning performance on the conservative side. We've made quite a few solid friends that way."

Results are a more compelling argument for the sale of a small car than exaggeration of miniature horsepower figures. And when these brilliant little engines are coupled with chassis and bodies that actually *contribute* to their effective urge, the results pour in. Except for a rare, super-costly semi-prototype, Porsches dominate their classes in competition. The fact of consistent superiority gets the job done in the sales room.

Now let's talk engines. Basically, the Porsche is an air-cooled flat four with overhead valves. There are two cylinders on each side of the crankcase, which is mounted on

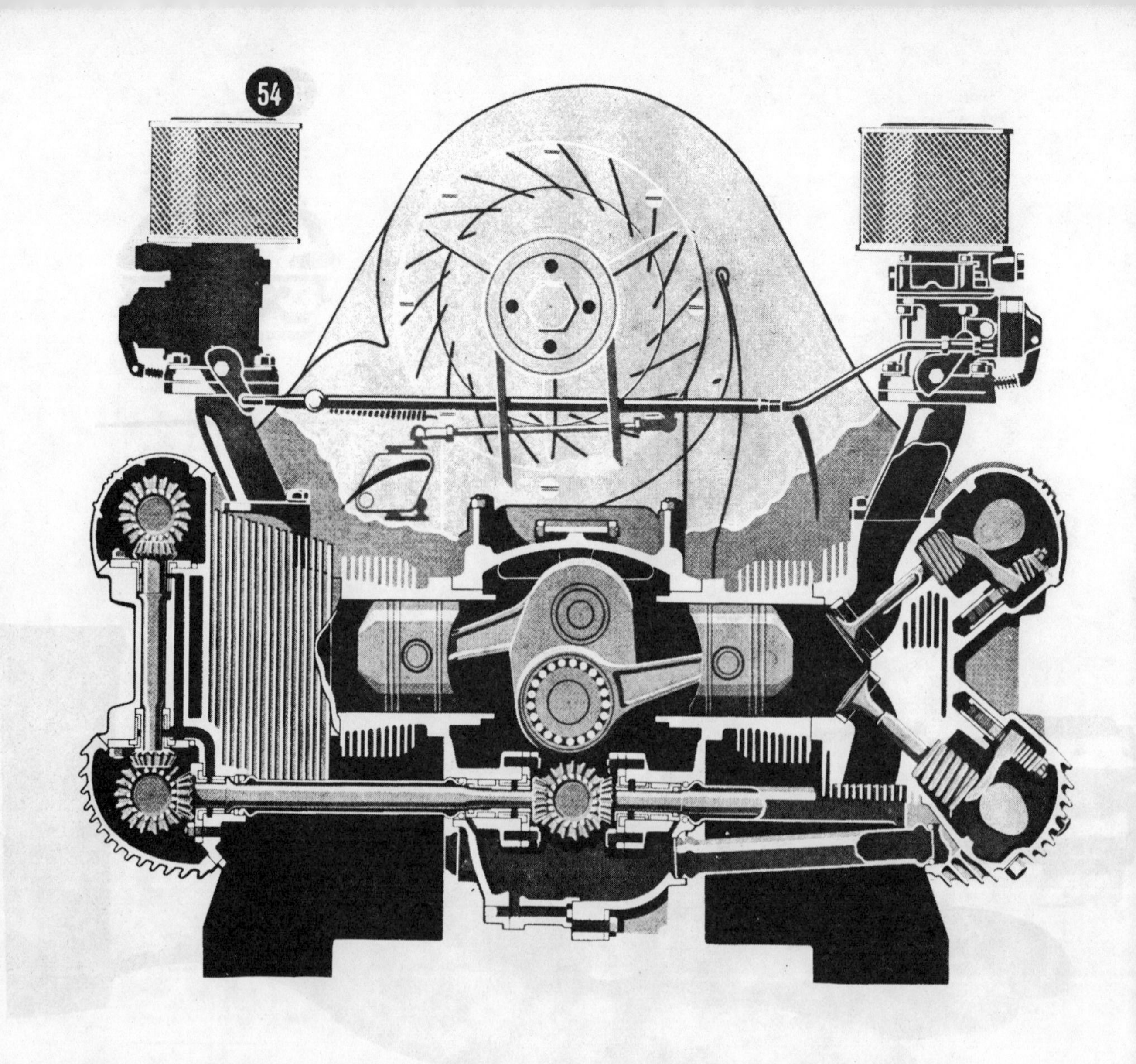

Cutaway showing working parts of dohc engine. Bevel gears drive camshafts. Con rods run on roller bearings at crank end. Note, domed pistons with oil ring below wrist pin.

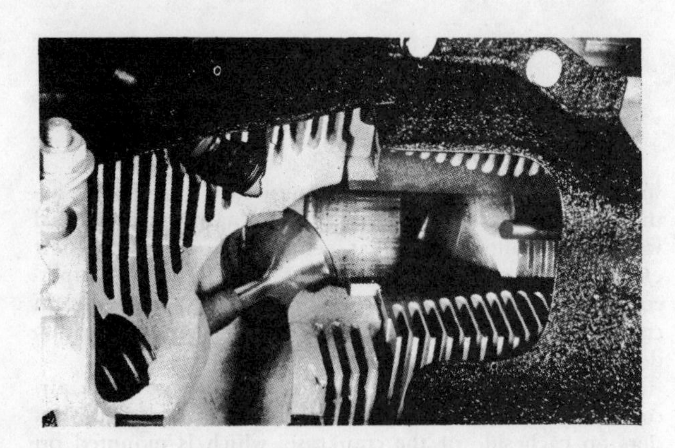

Partial cutaway shows intake valve, piston, and cylinder. Cylinders of all engines are of light alloy casting and do not use ferrous liners. Bores are chromium plated and lightly scored to hold oil film. Pistons are inverted-Vee type and do interfere with valves.

rubber blocks. Each engine is painstakingly hand-fitted and assembled and bears the initials of the skilled technician who performed the job.

The rigid, box-shaped crankcase is made of light alloy. It's in two pieces, split down the vertical center line. The main bearing bulkheads are narrow but massive and the mains are just under two inches in diameter — huge for such a small engine. There is a main bearing between each pair of crank throws, plus another main at the forward end of the crankshaft. Each of the main journals runs in solid, circular inserts except the No. 2 main, located between the front and rear cylinders. This runs in a split insert.

On older Porsche engines it was necessary to dismantle the crankcase in order to remove the camshaft. On the current engines this operation can be performed merely by removing the light-alloy cam-drive cover at the front of the case. If part of a crankcase is damaged both halves must be

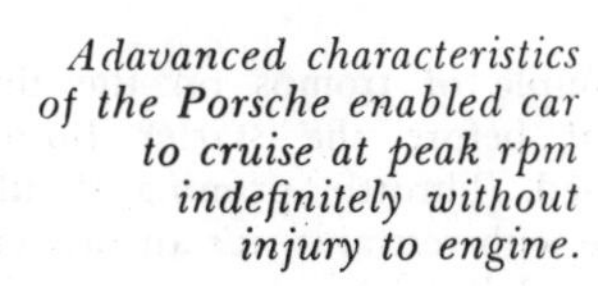

Adavanced characteristics of the Porsche enabled car to cruise at peak rpm indefinitely without injury to engine.

THE CURRENT PORSCHE ENGINES

	1100	1300	1300S	1600	1600S	1500GS (Carrera)	1500RS (Spyder)
Piston displacement	1086 cc 66 cu. in.	1290 cc 79 cu. in.	1290 cc 79 cu. in.	1582 cc 96.5 cu. in.	1582 cc 96.5 cu. in.	1498 cc 91.4 cu. in.	1498 cc 91.4 cu. in.
Bore & stroke	73.5x64mm 2.89x2.52 in.	74.5x74mm 2.93x2.91 in.	74.5x74mm 2.93x2.91 in.	82.5x74mm 3.25x2.91 in.	82.5x74mm 3.25x2.91 in.	85x66mm 3.35x2.60 in.	85x66mm 3.35x2.60 in.
BHP (SAE)	47 @ 4000	51 @ 4200	70 @ 5500	70 @ 4500	88 @ 5000	115 @ 6200	137 @ 6200
Torque (DIN), lb-ft	54 @ 3300	60 @ 2800	65 @ 3700	82 @ 2700	86 @ 3700	94 @ 5500	95.5 @ 5500
Compression ratio	7.0	6.5	8.2	7.5	8.5	8.7	9.5
Valve drive	Pushrod	Pushrod	Pushrod	Pushrod	Pushrod	DOHC	DOHC
Valve timing: In. opens BTC In. closes ABC Ex. opens BBC Ex. closes ATC	 2° 30′ 37° 30′ 37° 30′ 2° 30′	 2° 30′ 37° 30′ 37° 30′ 2° 30′	 19° 54° 54° 19°	 2° 30′ 37° 30′ 37° 30′ 2° 30′	 19° 54° 54° 19°	 38° 78° 78° 38°	 38° 78° 78° 38°
Valve diameter: Inlet Exhaust	 1.50 in. 1.22 in.	 1.50 in. 1.22 in.	 1.50 in. 1.22 in.	 1.50 in. 1.22 in.	 1.50 in. 1.22 in.	 1.875 1.5625	 1.875 1.5625
Carburetor throat diam.	1.26 in.	1.26 in.	1.26 in.	1.26 in.	1.58 in.	1.58-in. dual	1.58-in. dual
Spark plugs per cyl.	1	1	1	1	1	2	2
Connecting rod bearings	Plain	Plain	Roller	Plain	Roller	Roller	Roller
Lubrication	Conv.	Conv.	Conv.	Conv.	Conv.	Dry sump	Dry sump
Bore: stroke ratio	0.87	0.99	0.99	0.89	0.89	0.78	0.78
Piston speed @ max. bhp	1680	2040	2670	2180	2425	2690	2690
Bhp per sq. in. piston area	1.79	1.89	2.60	2.11	2.65	3.26	3.89
Bhp per cu.in.	.71	.65	.89	.73	.91	1.26	1.50

replaced; they come as fitted parts.

All con rod small ends contain bronze bushings. The 1100, 1300 use VW big-end inserts, and 1600 engines use lead-bronze big-end inserts which are interchangeable with the 1500. The 1300S, the 1600S, and the 1500RS and GS use normal main bearings but the rod big ends run on rollers. This is accomplished by means of built-up crankshafts and special, hand-finished, one-piece con rods. The roller-bearing shaft with a set of fitted rods packs a replacement price of about $513. In the case of either standard or Hirth roller crank, the flywheel is located by means of a single, central hollow bolt and eight locating dowels — an excellent, simple and handsome arrangement. The standard con rods are stubby H-section forgings with huge big ends. The bearing surfaces of all cranks are hardened.

The pistons in all Porsche engines are of the full-floating type, the wrist pins being secured by lock rings. Pistons for the various engines differ widely. All use two compression rings and one oil ring and these are arranged conventionally on the 1100 and 1300. The other engines use pistons on which the oil ring is mounted below the wrist pin and close to the bottom of the skirt. The 1100 piston has a domed crown with flyout clearance for the exhaust valve. The 1300's piston has a narrow, inverted-vee crown which requires no valve relief. The 1500 GS and RS use hairy, domed pistons with pronounced valve relief on two sides of the dome. The 1600 piston has a flat top and the pistons for the 1300S and 1600S have high flat tops with bevelled edges and exhaust valve relief.

The cylinders of all these engines are light alloy castings which do *not* use ferrous liners. Instead, the bores are chromium plated, then lightly scored to give a foothold to the oil film. This novel procedure is not as radical as it may

Four Cylinders

sound, having been developed and proved in German motorcycle practice. The hard chrome finish is amazingly wear resistant. I know of one Porsche owner who now has 82,000 miles on his engine and it still shows no sign of requiring a rebore. If it should reach that stage as it must eventually, he will not have to invest in new cylinders. Lockheed will rechrome the original ones for a modest price. Each Porsche cylinder, new or replacement, comes with its own piston, tailored to fit. The sloppy clearances popularly associated with air-cooled engines don't apply to the Porsche. The 1300 and 1300S, for example, use pistons that are only six-tenths of a thousandth of an inch smaller than their cylinder bores.

Each bank of two cylinders carries a common, removable light-alloy head which, like the barrels, is deeply finned for optimum cooling. The head contains steel or bronze valve seats, valve guides and spark plug bosses. The cylinders are spigoted to permit the heads to fit down around them and no head gaskets are required.

The overhead valves of all but the 1500GS and RS are pushrod operated and are arranged in a wedge-shaped variation on the vee-inclined theme. The camshaft rides in three bearings and is driven by a light-alloy helical gear. With admirable simplicity, each lobe on the camshaft alternately operates one valve of two opposed cylinders, by means of pushrods and rocker arms. Mushroom-type cam followers act on the pushrods, which are of light alloy with a steel center section. This bimetal combination has a coefficient of expansion practically identical with that of the head and block assembly. Valve lash is adjusted conventionally at the rocker arms.

The 1500 GS and RS are full-race versions of the foregoing engines, as a glance at the table shows. The heads of these engines have true hemispherical combustion chambers, classically vee-inclined valves, two spark plugs per cylinder, and dual overhead camshafts. That adds up to four camshafts for the engine.

These cams are driven by shafts and bevel gears. Instead of using conventional, somewhat clumsy Oldham-type couplings, the main cam-drive shafts are jointed by means of male and female splines. These permit changes, including compression ratio and cylinder height, to be made with no modification of the cam drive.

The lobes on these d.o. cams are enough to quicken the pulse of any roller tappet devotee. They're as blunt as the end of a broom handle and act upon radiused finger-type cam followers.

The followers are heavily spring-loaded and ride on adjusting buttons placed over the valve stems. The exhaust valves are sodium cooled and their heads are higher at the center than at the periphery so that, if a piston should hit a valve, it will close it, not bend it. Double-nested valve springs are used in all the Porsche engines.

The cylinders and heads are cooled by means of ducted air that is force-fed by a powerful fan driven by an adjustable v-belt pulley on the crankshaft. An oil cooler also is mounted in the path of cool air flow and all lubricant must pass through it before it reaches the engine's bearing surfaces. Correct viscosity of the oil is maintained in cold weather by means of a by-pass valve ahead of the cooler.

All oil passages in the crankcase, blocks and heads are lined with copper tubing. The lube system delivers pressure oil to all friction surfaces with the exception of cylinder walls and wrist pins, which are splash-lubed. On the 1500GS and RS engines a single pump handles both oil feed and scavenging. In addition to a conventional cartridge-type oil filter the system includes a large "magnetic filter" in the bottom of the sump for extracting ferrous particles from the oil.

The Porsches use a single-plate dry clutch built into the flywheel; the clutch pressure-bearing requires no maintenance. If there is an Achille's heel to these engines, the clutch is it. This is not necessarily because of any defect inherent in the robust 7.13-inch clutch but is more a function of driving conditions and techniques. Protracted drag racing from traffic light takes its toll and changes a delightfully soft and easy clutch into a rough one. The Porsche is built to tolerate hard use but leadfoot owners will do themselves a favor by not expecting the clutch to live out its life constantly slamming into a flywheel spinning at 4000 revs or more.

The pair of Solex carbs fitted to Porsches have generous acceleration pumps. The engine can be brought to life on the coldest morning with just a couple of tromps on the throttle pedal before the starter button is pressed. Throttle priming should be done with restraint; it's an easy engine to flood if you're careless.

Although the two carbs are nearly a yard apart, the linkage that joins them is tight, positive and substantial. All engines but the hot 1500's have single-throat carbs. The GS and RS come with Solex-twin-throat units and are often fitted with twin-throat Webers for competition; with linkage and manifolding they cost about $900. The Webers make little difference in ultimate power output but they do make for a marked improvement in acceleration and prompt, smoothly continuous throttle response.

Dual ignition on the GS and RS engines is credited with increasing their total power output by better than ten percent. One distributor feeds the juice to one plug in each cylinder. Thus, if one distributor should fail, you can stay in the fray on the remaining, separate ignition system.

When laying down the specifications for the original VW engine, Dr. Porsche chose air cooling largely because of its indifference to climate: "air neither freezes nor boils." Air cooling was a must for the projected universal car that could be left in the open in any climate, but that would always be reliably ready to go.

There are other advantages to air cooling, including nicer control in the casting process and elimination of the costly, heavy, superfluous bulk of water jacketing and radiator. The one disadvantage seems to be noise: a water jacket around an engine is a highly effective sound deadener. But the problem of noise has been adequately coped with in latter-day production of the Porsche family of engines, as far as the Porsche-VW market is concerned. If customers should ever begin to complain seriously, it could be coped with even more. Aside from this one objection, everything is in favor of retiring the water jacket along with the whip socket.

Dr. Porsche's *boxer motoren* are a revolution in power plant design, a sort of ultimate simplification. The basic design is a modern, intensely rational solution to the problem of propelling a car by means of a piston engine. It probably will retain its excellence as an answer to this problem as long as piston engines are being made. #

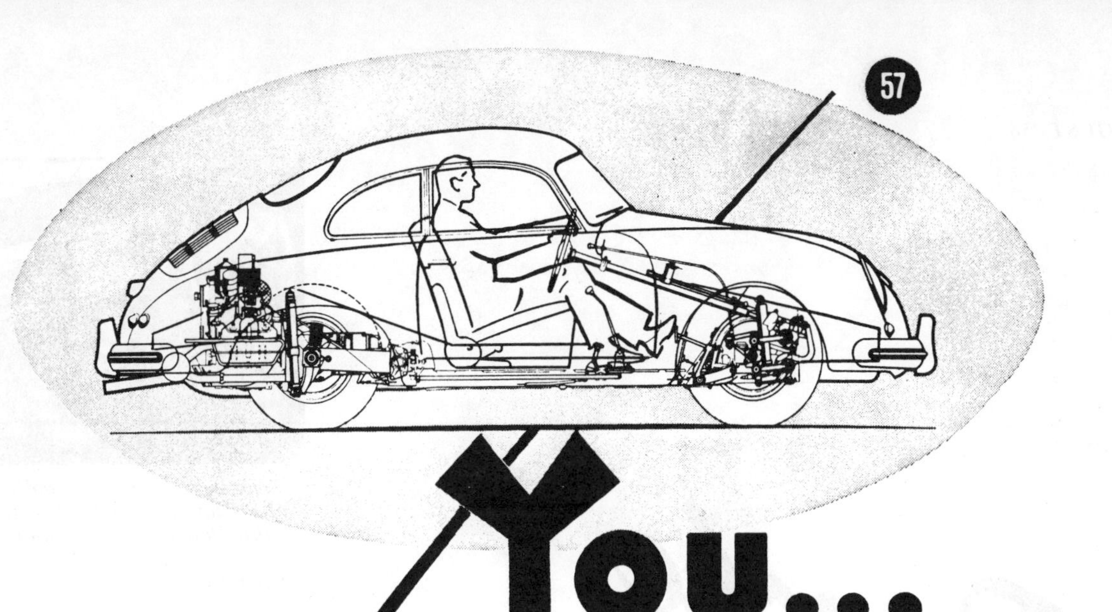

Two way short wave radio is part of the latest equipment of the Autobahn Police. Switches and dials of unit set neatly in the glove compartment.

Porsche prowler

Equipped with the necessary accessories for police work, the Porsche takes on a forbidding look. Dual horns mounted on bumpers are high pitched and blare in an alternating two-toned sound. On fender is powerful blue emergency light.

By JESSE ALEXANDER

HIGHWAY conditions in West Germany have gone from bad to worse since the end of the war. A booming German automobile industry is pouring out huge amounts of new cars every month that are immediately gobbled up by a flush and auto-starved population. But on the Autobahns traffic conditions are especially bad — overcrowded roads, reckless drivers, autobahn bandits, and uneducated truck drivers have made this terrific highway system into a nightmare which is often referred to by both US Army MP's and German police as the most accident prone stretch of road anywhere in Europe.

Conditions have made at least one German state re-evaluate their rolling stock. North Rhine Westphalia's autobahn police are giving up their three and four year old Mercedes

Extra generator mounted at left of regular generator supplies auxiliary batteries with power for short wave set.

220's and poky beat-up Volkswagens for newer and faster equipment — namely the beautifully outfitted unit pictured on these pages; a white 1600 Type 356A Porsche convertible equipped with two-way radio. The cars will be manned by specially picked, extra tough officers who will sport white crash helmets and green uniforms.

The radio installation is especially neat; as can be seen, it is fitted directly behind the seats with two extra batteries charged through a second generator installed in the engine compartment. Dual high-pitched horns are mounted on the front bumpers and they give out an alternating two-tone sound that is enough to lift the hackles of every 300 SL this side of the Rhine. A powerful blue emergency light is mounted on the left front fender. The radio installation is by Schaub-Lorenz.

Power is supplied by the well-known 70 bhp 1600 "normal" engine giving the police cars a top speed of over 100 miles per hour and enough acceleration to take on any modified Volkswagens who get fresh. It's interesting to note that the German police are almost entirely equipped with open vehicles. Even the huge riot trucks fitted out with floodlights and rescue equipment have soft tops — and a common sight on the Autobahn is an elderly dark green VW convertible with a tall white-capped polizist, waving his arms directing traffic. The authorities apparently prefer the extra visibility and easy access afforded by a convertible to four-door sedans.

These new highway police with their fleet of Porsches will present a picture formidable enough to put the fear of God into most all would-be law breakers, at least in North Rhine Westphalia. And the idea *could* get contagious! Look in your rear view mirror. #

Outside of the additional equipment mostly located inside, the Porsche remains unchanged in body contour and paint.

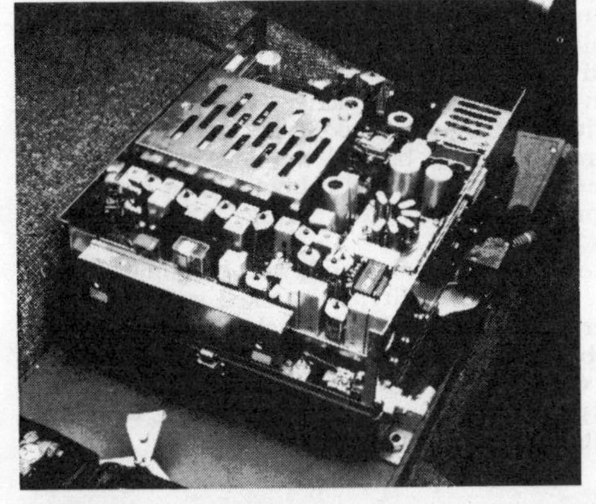

Combined transmitter and receiver unit sits behind driver and assistant. Two six volt batteries connected in series power the short wave set.

A fully curved screen, together with slim pillars, gives excellent forward visibility. The body by Reutter is wind cheating, and even at high speeds there is practically no wind noise

The Autocar ROAD TESTS 1704

Porsche 1600

IN times of increasing standardization and sameness, a car of unusual design and appearance naturally arouses interest, and especially so when, in its individuality, it has proved successful. The Porsche has become accepted as something rather special, and ownership of one of these cars is taken as an indication of a keen and discriminating driver. Much has been said and written about the rear-engine layout, and its effects on handling, but the fact remains that in races and rallies Porsches have done extremely well against much more powerful opposition.

Based originally on Volkswagen components, the first Porsche appeared in 1949, subsequently developing into a separate entity with Ferry Porsche, son of the founder, at the head of the factory in Stuttgart. Like the Volkswagen, the Porsche still retains an air-cooled, rear-mounted engine, with pushrod operation of the overhead valves, and the rear suspension consisting of swinging half-axles and torsion bars.

The car tested is the latest Type 356A 1600—the Porsche *Damen*, or, literally, the "Ladies' Porsche". It is the most docile model of a line of production cars which has, at the high-performance end, the 125 b.h.p. G.T. Carrera. The 1600 model now constitutes the bulk of Porsche production, and its very similar predecessor was tested in *The Autocar* of 4 May 1956. Since then, the main change has been the fitting of Zenith twin-choke downdraught carburettors in place of the original single-choke units. Though this has not made much difference to the out-and-out top speed, the engine has become much more tractable, and acceleration figures have been improved throughout the range.

First impression of the car is the manner in which the Competition Department has influenced its conception; and the feeling, as soon as one takes the driving seat, that it is a vehicle with inbuilt character. The driving position is very comfortable, the quality of finish is first class, and attention to detail is above average.

For example, the hand throttle on the dashboard may be set in any one of an almost infinite number of positions; like all other controls, it functions easily and precisely, with no lost movement. There are elastic straps within the lid of the glove locker, to hold pencils, average speed computers and other oddments; and the sun vizors, instead of being made of unyielding material, are of foam rubber, leathercloth covered, and would not inflict injury in the event of an accident. These and many other small points give the impression that a great deal of thought has been put into the car by people who have experience of competition driving—or who get a joy out of their motoring.

The Reutter seats are outstandingly good, the backrests being adjustable from the vertical almost to a prone position; upholstery is comfortable yet firm and gives ample support to the back and shoulders as well as the thighs; the seats also locate the occupants laterally. With variable rake of the backrests, the fore-and-aft adjustment of the seats (which is considerable), and the variation of reach available even on the pedal pads themselves, it is possible to set the driving position to suit anyone. The "bonnet" (front boot lid) falls away quickly; all but the smallest drivers can see over the scuttle to the road a few feet ahead of the car, and both wings are visible from the driving seat. Upholstery is in good-quality leathercloth, with floor coverings in rubber; a form of carpeting is used to cover the interior of the body sides.

There is a fair amount of overhang—at the sides as well as at the back and front—and a few miles of traffic driving are needed to learn to place the car exactly. The side overhang can be deceptive when parking alongside high kerbs, the body panels projecting well beyond the nave plates of the wheels. There are sensibly placed rubbing strips along the lower edge to protect the body from damage inflicted in this way.

The windscreen is fully curved and, with the slim pillars, gives excellent forward visibility; it is raked sufficiently to allow the driver to see traffic lights from close beneath them. All-round visibility is satisfactory, though, by the latest standards, the windows are a little on the shallow side; the driving mirror gives a full view of the road astern, with the minimum of blind spots at the quarters.

Though the rear window is shallow, it does not restrict the view of the road astern in the mirror. Exhausts are led out through the lower ends of the over-riders, and a rubbing strip along the lower edge of the body sides protects the panelling against damage by high kerbs

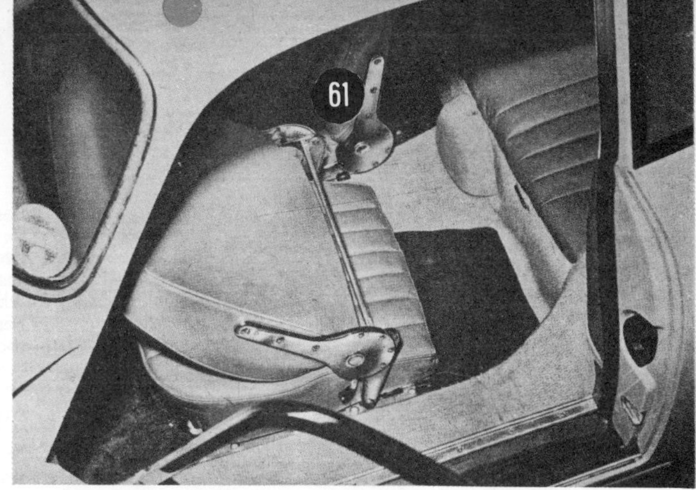

Instruments, grouped in front of the driver, are easily read. The backrests of the seats are fully adjustable for rake—almost to a prone position—by means of the small lever seen at the pivoting point; the seats themselves are comfortable and give good support. Limited though not uncomfortable space is available for two passengers in the rear, occasional seats

The hand-brake is of the under-scuttle, pull-and-twist type; though strong and efficient (it will hold the car on a one-in-three gradient in both directions), it is a little out of keeping with the character of the Porsche; a pull-up lever beside the driving seat would be more convenient. The pedals are a little too close together, and the accelerator is so arranged that its hinged, lower edge is too close to the brake pedal when the latter is depressed. This, together with the offset towards the centre, makes for awkwardness, at least on early acquaintance. The accelerator linkage is smooth and progressive—quite an achievement when the pedal is so far from the carburettors.

Instruments include a speedometer and a revolution counter, and fuel and oil temperature gauges combined in a third dial. In the composite dial are included ignition and oil pressure warning lights and, in the rev-counter, the main beam tell-tale and indicator repeater lights. An ordinary push-pull switch which operates the lamps is slightly awkwardly placed beneath the steering wheel; there is a combination switch for the ignition and starter. The various heater-ventilator controls, in combination with the trailing quarter windows, make it possible to control the interior temperature over a very wide range, and there are no traces of fumes when the heater is working. A defective fuse can be identified and replaced without getting out of the car, for the fuse box is located beneath the scuttle. The fuel tap, of substantial proportions and marked "On," "Off," and "Reserve," also beneath the scuttle, is operated from the driving seat.

Provided the instruction to depress the throttle pedal fully three times before operating the solenoid is obeyed, the engine starts easily, even after a night in the open. In this respect, incidentally, Porsche owners do not have to worry about sub-zero temperatures at night; as the engine is air-cooled, there is no water to freeze. The engine runs smoothly and quietly, and produces the characteristic slight thrumming note of the flat-four. Though pleasantly light in operation, the clutch on the car tested was a little indecisive, and was inclined to slip mildly until readjusted.

As soon as one moves off from rest, it is clear that the Porsche's individual character is not confined to its design alone. One quickly begins to feel a part of it—as though one were wearing, rather than sitting in the car. Acceleration is impressive, particularly so because of the seemingly effortless way in which the car gathers speed, the almost complete absence of transmission noise, and the low level of road induced body noise. The gear change is smooth, and the lever easily reached, but it requires a fairly long movement; the synchromesh is good, though it is possible to override it in snap changes upwards from first to second. In contrast with the present tendency to make first an emergency gear for getting away on steep gradients, the Porsche's lowest ratio is eminently usable, giving a top speed of 25 m.p.h.; in traffic, one soon appreciates the unusual feature, with a four-speed box, of synchromesh on this ratio.

Somehow the car seems to slip through traffic, eagerly taking advantage of every gap that presents itself; the impressive performance in first and second in part accounts for this, and the exhaust note under full acceleration is not enough to make bystanders stare. Lower ratios need to be used freely in traffic, and one seldom uses top gear in town—though with the twin-choke carburettors it is possible to trickle along at 20 m.p.h. in top, and it is easy to forget that one is still in third when the derestriction signs come up. Acceleration in third is maintained at a remarkably constant rate from 20 to 60 m.p.h. which helps greatly when overtaking on crowded British roads.

It is on the open road that the Porsche really comes into its own, and one begins to wonder how it ever came to be dubbed the *Damen.* As unobtrusively as it goes about its business in heavy traffic, it gobbles up the miles of main highway amazingly quickly. The sole indication—so far as handling is concerned—of the combination of rear engine and swing-axles is a slight outward "lurch" of the rear when the car begins to change direction on entering a corner; this is particularly noticeable on a fast road with a succession of bends in opposite directions, when the movement becomes more accentuated. It is in no way worrying, however, and one soon grows accustomed to it; after this initial suggestion of oversteer—if indeed it can be called oversteer—the Porsche handles normally through a corner.

Like all cars with a good power-weight ratio, it can be made to break away at the rear—more particularly in the wet, of course—but it does so in a perfectly controllable manner. There is no question of a vicious whip-round of the tail—in the dry or wet—and the power can (even should) be kept on during recovery.

Steering is light, positive, precise and quick, with 2½ turns from lock to lock, and the car will maintain a straight course at speed. In fact, a delicate touch on the wheel is desirable; if one grips rather than "floats" the wheel, as with all cars with light and comparatively direct steering, the Porsche tends to suffer from over-correction, and to wander. The somewhat solid wheel seems strangely out of keeping with the delicacy and precision of the handling, and one feels

Tucked away at the back and occupying remarkably little space, the flat-four engine will propel the car at 103 m.p.h., and is itself reasonably accessible. Auxiliaries are reached easily, though it is necessary to jack up the rear of the car to adjust the valve clearances

Sealing is of a high standard, and unless the trailing, rear quarter vents are opened the heating-ventilating system has little effect

that a light alloy, flexible wheel with thin wooden rim would enhance the already pleasant feel of the car.

Flexibility of the 1600 engine has been much improved. The rev-counter is marked in green from 2,500 to 4,500 r.p.m., which indicates the speed range suitable for wide throttle openings. Though the car will run smoothly, with no snatch, at 20 m.p.h. (roughly 950 r.p.m.) in top, it does not like pulling away from this speed on an open throttle. Between 4,500 and 5,000 r.p.m. on the standard 1600 (5,000 to 5,500 on the Super) the rev-counter is marked in red, 5,000 r.p.m. giving an indicated 110 m.p.h. (103 m.p.h. true). On a level road the engine will reach this speed in top gear—in fact, given time, it will exceed it. After several miles of a sustained 90 to 100 m.p.h. the oil temperature did not rise significantly on a warm day, and there was no increase in mechanical noise when the car was brought down to normal speeds again.

An indicated 90 m.p.h. cruising speed—or even faster—seems well within the car's capabilities, and is achieved with surprisingly little wind noise provided, of course, that side windows are kept closed. When they are open there is a curious pulsation of the air within the car, which becomes painful to the ears.

The suspension is relatively stiff over small bumps, but remarkably soft over larger undulations—yet there is no appreciable roll on corners. In conjunction with the first-class seats, this gives a very good ride; after a fast run of seven hours or so, one feels neither stiff nor tired.

Brakes of the car tested proved deceptively powerful at light pedal pressures; until one grew accustomed to this characteristic there was a tendency to lock the wheels. Though there were no signs of fade, the pedal travel increased appreciably during the period when repeated stops from high speed were necessary as the performance figures were being taken. Even in the wet, the car normally stops quickly and evenly; with heavy braking on certain surfaces, there is early locking of the rear wheels, and retardation is then much reduced.

Basically a two-seater, the 1600 will take one—even two at a pinch—passengers in the well upholstered back seat for short journeys; lack of legroom, rather than lack of comfort, is the limiting factor. For touring, however, it is no more than a two-seater, as luggage space is obtained by folding flat the backrest of the rear, occasional seat; this can scarcely be regarded as extra luggage space—it is very considerable—because, under the "bonnet," there is room for little more than an overnight bag. In the nose are housed the spare wheel, fuel tank, tools, jack and battery (in a compartment beneath the spare wheel, which has to be removed before the battery acid level can be checked). The under-bonnet compartment is lined in rubber, and such luggage as can be accommodated lies out of harm's way between the spare wheel and fuel tank.

In keeping with the quality of finish of the remainder of the car is the tool-kit; contained in a smart leather-cloth roll are a set of chrome-vanadium spanners, dial-type tyre pressure gauge, spare sparking plug, fan-belt and other necessary equipment—together with a tin of retouching enamel for the coachwork.

Perhaps one of the most remarkable things about the Porsche is the economical way in which it achieves its performance. During the test period, which included the high-speed runs and successive bursts of full acceleration when the performance figures were being taken, together with a great deal of driving in London, the car showed an overall fuel consumption of 32 m.p.g. Clearly, this figure would be improved considerably in normal, open road conditions.

The fuel tank, which holds 11½ gallons, has a sensibly large, flush filler opening directly into the tank; thus it will take the full flow of the most powerful electric pump without any risk of blow-back. With an average fuel consumption of well over 30 m.p.g., this tank capacity gives a useful cruising range of at least 350 miles; there is a reserve (included in the 11½ gallons) of one gallon.

As befits the sort of car that appeals to the owner who will do much of the servicing himself, the engine is reasonably accessible. Despite the compact dimensions of the compartment that houses them, the auxiliaries, too, are easy to reach—coil, distributor, fuel pump, Fram oil-filter and so on. The forward pair of plugs are somewhat tucked away and, without the special spanner provided in the tool-kit, would be difficult to reach. To set the valve clearances it is necessary to jack up the back of the car—or, better still, use a hoist—when, from beneath, the valve covers are easily removed.

Both the front "bonnet" and rear engine compartment catches are released from within the car by means of pull-knobs. In the event of failure or breakage of these wires, there are alternative methods of opening the two compartments, which are explained in the very full and detailed instruction book, and which would be wellnigh impossible for a thief to discover. Both doors can be locked either from within the body itself, or from outside with the ignition key.

Even more powerful head lamps would be desirable for drivers intending to use the full speed of the car at night, though for cruising speeds of from 70 to 80 m.p.h. they are quite adequate. The central button on the steering wheel flashes the head lamps in daytime—and serves as a greeting between Porsche drivers, it appears—and, when the lamps

Left: Backrest of the rear, occasional seats folds flat to provide considerable luggage space—in what is, then, a two-seater. The height of the rear window permits a substantial depth of luggage to be stowed before it obstructs the view in the mirror. Right: The 11½-gallon fuel tank has a sensibly large filler. Luggage space within the rubber-lined front compartment is limited; a fitted suitcase would make best use of the space available

Porsche 1600 . . .

are in the dipped position at night, the same button brings on the main filaments. The horn, scarcely loud enough for so fast a car, is operated by a full ring on the steering wheel; the dipswitch is foot operated by a plunger, near the clutch pedal, that is very similar to the one that operates the screen-wash. Rheostat-controlled by twisting the main lamp switch, the instrument lights throw no reflections on the screen at night, and, at full low, do not distract the driver, though the instruments are still legible.

The Porsche is a car with a personality of its own. Very compact, it is ideal in heavy traffic, being tractable, quiet and quick off the mark; yet on the open road there are few cars that can catch a well-driven Porsche, whatever their size; in the mountains there are even fewer.

With a kerb weight of 16¾ cwt, of which about 59 per cent is on the rear wheels, the car is definitely tail-heavy by front-engine standards; figures for a comparable front-engine car are 55.6 per cent front, 44.4 per cent rear. Yet years of development have gone far towards removing any handling shortcomings resulting from the rear engine layout; once a driver has grown accustomed to the car, it is easily controllable on dry and wet roads, provided, of course, as with any high-performance car, that the power-weight ratio is borne in mind, and undue liberties are not taken. The standard of finish is excellent throughout, and meticulous attention to detail is evident.

PORSCHE 1600

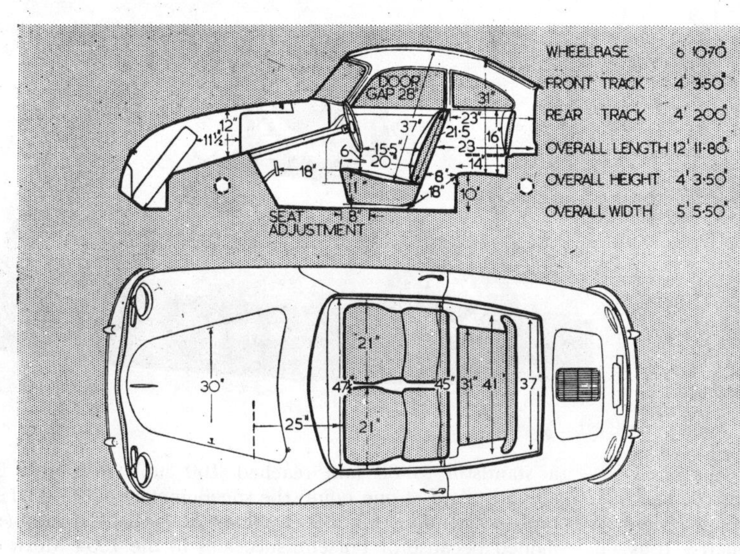

Scale ¼in to 1ft. Driving seat in central position. Cushions uncompressed.

PERFORMANCE

ACCELERATION TIMES:

Speed Range, Gear Ratios and Time in Sec.

M.P.H.	3.92 to 1	5.44 to 1	7.81 to 1	14.08 to 1
10—30..	—	—	5.0	—
20—40..	—	7.2	4.5	—
30—50..	11.1	7.3	—	—
40—60..	11.2	7.1	—	—
50—70..	12.7	8.4	—	—
60—80..	14.4	11.8	—	—
70—90..	20.3	—	—	—

From rest through gears to:

M.P.H.	sec.
30	4.1
40	6.2
50	9.5
60	14.1
70	18.1
80	25.5

Standing quarter mile 19.1 sec.

MAXIMUM SPEEDS IN GEARS:

Gear		M.P.H.	K.P.H.
Top ..	(mean)	102.4	164.5
	(best)	103.0	165.8
3rd	..	80.0	128.7
2nd ..	.. ..	50.0	80.5
1st ..	.. ..	25.0	40.2

TRACTIVE EFFORT:

	Pull (lb per ton)	Equivalent gradient
Top	210	1 in 10.6
Third	301	1 in 7.4
Second	460	1 in 4.8

BRAKES (at 30 m.p.h. in neutral):

Pedal load in lb	Retardation	Equivalent stopping distance in ft
25	0.31g	98
50	0.77g	38
75	0.97g	31

FUEL CONSUMPTION:

M.P.G. at steady speeds

M.P.H.	Direct Top
30	74.8
40	62.5
50	51.9
60	43.8
70	37.8
80	32.0
90	26.8

Overall fuel consumption for 850 miles: 31.8 m.p.g. (8.8 litres per 100 km).
Approximate normal range 29-38 m.p.g. (9.7-7.4 litres per 100 km).
Fuel: Premium grade.

TEST CONDITIONS: Weather: Dry, no wind. Air temperature 68 deg F.
Acceleration figures are the means of several runs in opposite directions.
Tractive effort obtained by Tapley meter.

SPEEDOMETER CORRECTION: M.P.H.

Car speedometer	10	20	30	40	50	60	70	80	90	100	105	110
True speed ..	10	19	28	37.5	47	56	65	74	83.5	93	98	103

DATA

PRICE (basic), with fixed-head coupé body, **£1,330.**
British purchase tax, **£666 7s.**
Total (in Great Britain), **£1,996 7s.**
Extras: Radio to choice.

ENGINE: Capacity, 1,582 c.c. (96.5 cu in).
Number of cylinders, 4, horizontally opposed.
Bore and stroke, 82.5 × 74 mm (3.25 × 2.91in).
Valve gear, o.h.v., pushrods.
Compression ratio, 7.5 to 1.
B.H.P. 70 (S.A.E.) at 4,500 r.p.m. (B.H.P. per ton laden 70.7).
Torque, 82.2 lb ft at 2,800 r.p.m.
M.P.H. per 1,000 r.p.m. in top gear, 22.0.

WEIGHT: (with 5 gals fuel) 16.8 cwt (1,883 lb).
Weight distribution (per cent): F, 41.3; R, 58.7.
Laden as tested, 19.8 cwt (2,219 lb).
Lb per c.c. (laden): 1.4.

BRAKES:
Method of operation, hydraulic.
Drum dimensions: F, 11in diameter; 1.57in wide. R, 11in diameter; 1.57in wide.
Lining area: F, 61 sq in. R, 61 sq in. (123.1 sq in per ton laden.)

TYRES: 5.60—15in.
Pressures (lb sq in): F, 22; R, 24 (normal).

TANK CAPACITY: 11.5 Imperial gallons.
Oil sump: 9 pints.
Cooling system: air cooled by fan.

STEERING: Turning circle:
Between kerbs, 32ft 11in (L); 31ft 10in (R).
Between walls, 34ft 11in (L); 33ft 8in (R).
Turns of steering wheel from lock to lock 2.5.

DIMENSIONS: Wheelbase, 6ft 10.7in.
Track: F, 4ft 3.5in; R, 4ft 2in.
Length (overall): 12ft 11.8in.
Width: 5ft. 5.5in.
Height: 4ft 3.5in.
Ground clearance: 6.5in.

ELECTRICAL SYSTEM: 6-volt; 84 ampère-hour battery.
Head lights: Double dip; 35-40 watt bulbs.

SUSPENSION: Front, independent, trailing arms and torsion bars. Rear, independent, swing axles and torsion bars.

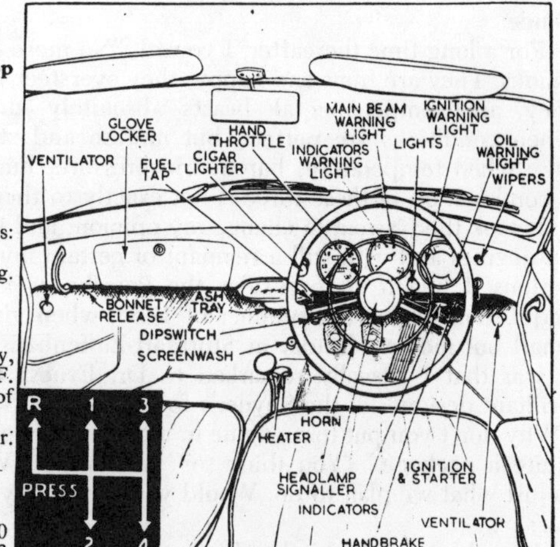

The Porsche GT exhibits none of the undesirable handling features of earlier models as it screams around loop at Thompson.

PORSCHE CARRERA

The back wheels transmit 128 hairy horses to the road as the test Carrera gets off the mark with Bentley and Porsche factory representative aboard.

THE illustrious firm of Porsche has turned out many models since its original export hit these shores back in 1951; and for each of these models there have also been rafts of road tests, features and comparison charts. It is claimed that magazine road tests are impartial because the writer has no axe to grind—yet what writer could be more objectively critical than an owner intimately acquainted with his car and dedicated to extracting from it the last ounce of performance?

I have owned six Porsches since 1952, beginning with the successful 356 Coupe which earned me a number of trophies in races and hillclimbs, and ending with an improved version of this particular type, in which—at the close of 1953—I spun out at 90 mph during an airport race and then gyrated backwards off the course. That I did not kill myself was due neither to the Porsche, nor (as many thought) to my driving skill; but solely to benign Providence.

For a long time thereafter, I vowed, "No more Porsches, thanks. They are front-end happy; they oversteer like mad; they are temperamental beasts—absolutely unbeatable when conditions are perfect, but morose and stubbornly lazy when temperature, barometric pressure, humidity or a combination of these three is not exactly to their liking."

It took the Carrera to change my opinion, and then only by degrees and with still a remnant of certain minor reservations. I began to reconsider the Porsche as a possible acquisition during the summer of 1955, when visiting the small but modern factory at Stuttgart-Zuffenhausen. Then it was that I casually remarked to Dr. Ernest Fuhrman, brilliant designer of the Spyder's four-cam 1500GS engine, "Why don't you put that engine in a coupe? It would make quite a package." "You think so?" he smiled. "Well, that is just what we plan to do. Would you care to try the pilot car?"

I would and I did; and as I expected, it was a revelation in performance. For instance, we broke nine seconds from a standstill to 60 and reached 100 mph in under 20 seconds. And at one point, the speedometer registered 125 mph—astounding for a 1 1/2-liter car. The Carrera, (so named because of Porsche successes in the 1954 Mexican road race,) developed 100 (DIN) bhp @ 6200 rpm, or 10 bhp less than the racing Spyder, and serves essentially as what Dr. Fuhrman termed as, "a detuned version providing extra performance for high speed, Gran Turismo competition, with more power than the pushrod engine could produce." DIN, by the way, stands for *Deutsche Industrie Norm*, a basic German power output measure more critical than SAE standards. In this rating, the power consumed by all accessories is taken into account, so arriving at a true figure which would obviously not be the one selected by the ad copy boys extolling Detroit products.

All very impressive, but still this prototype machine did not exhibit markedly improved handling characteristics, for the simple reason that the Carrera engine had been dropped into a 356 chassis, and not into the 356A which made its bow at the Frankfort Show a month later, on September 22, 1955.

The 356A chassis, as you probably know, was something else again. Front end geometry was revamped to give five degrees of caster while the steering was fitted with a more substantially anchored stabilizer bar and a hydraulic damper. Front torsion springs were made softer and slightly heavier to lower the spring rate and permit additional up-and-down wheel travel; the rear shock absorbers were relocated further back and the floor line was lowered 1⅜-inches to facilitate entry in the coupes.

As a result the Porsche acquired a brand-new handling and riding personality. If the steering lost its featherlightness, it also shed that alarming front-end tendency to "float" and wander at high speed. Then too, the traditional oversteer gave way to neutral steering; and although the car offered a slightly "softer" ride than before, it became as tractable as a front-engined machine and could be

How to make a standard Porsche owner green with envy... take him for a ride in a Carrera Gran Turismo Speedster

Gran Turismo Speedster

Sturdy black rollbar contrasts strangely with the other finely wrought cockpit appointments of meticulously finished Porsche sports-racer.

"Tops are for touring" seems to be the motto of the Stuttgart firm. Porsche top is easy to put up or down, but might prove confining on long trip.

driven in much the same way.

The logical sequence was the Carrera Speedster, favored by US amateur race drivers in 1956 and 1957, and produced in ample numbers to gain acceptance as a production model. All this I observed with interest and curiosity, yet still I hesitated to wade in and get my feet wet. After all, a Carrera is a hefty investment—they want *mucho dinaro* for so small a package—and before you mortgage the homestead and water the baby's formula, you have to think twice.

But as the months passed and owners spoke glowingly of their acquisitions, and (best of all) I drove the Carrera Speedster, I became convinced of one thing: any competition driver searching for the true all-purpose sports car—one that you could drive to the A & P and then race over the weekend—need not bother to look any further than the Carrera. Its enormous margin of strength and overall robustness, combined with reasonable docility and truly practical bodywork, seemed to supply all the answers.

Then, last year, when the lightened and more powerful GT (Gran Turismo) Carrera Speedster made its appearance, I knew this was for me. This model, which has only been changed in detail since its inception, was given the Spyder's front brakes, providing an increase in brake lining area from 115 to 128 sq. in.; it also borrowed the Spyder's engine practically intact (15 more DIN bhp), except for a slight decompression to 9 to 1 and a less radical exhaust system. In addition, the GT Carrera was given the Spyder's worm gear steering and was shorn of heater and undercoating. The loss of these impediments, together with the substitution of aluminum for the doors, hood and engine lid, and the use of lighter bucket seats, effected a weight saving of over 140 pounds. Parenthetically, the gas tank was enlarged.

When I finally took delivery of a metallic silver-gray Speedster with black leather, I was almost too late. My acquisition was the first of a batch of four, all of which were instantly sold out, and since then there has been only one more car and *finito!* No more GT Carreras until February 1959 at the earliest. Actually, I gained a little by waiting, because my car has almost everything of the very latest: a new V-drive for the twin distributors, taken off the crankshaft, instead of the individual cam boxes; a so-called ram-type air intake featuring louvers in the engine compartment lid; built-in rollbar posts and Fren-do competition type brake linings. But even that was not quite the end of it. The three cars immediately following also had those fabulous Koni adjustable hydraulic shock absorbers as standard equipment, whereas I had to purchase a set—and they were not cheap, although the greatest thing since shocks were first thought of.

We did not immediately come to terms, the GT Carrera and I. After the first romantic flush of ownership was over and my friends had exhausted their envious "Oh's!" and "Ah's!", the car seemed to lose all its pep and would not pull the skin off a rice pudding in any gear. This was poor reward for meticulously breaking-in the engine during the first 750 miles. The manual said that 4500 rpm must not be exceeded for 600 miles; then 5000 rpm for the next 2400 miles. I proceeded according to the manual, but the plugs oiled-up and performance fell way below that of a decently tuned 1600 standard pushrod Speedster. This was so galling that I nearly threw in the sponge. I am glad that I stuck with the car. The rings had not seated properly (a rare occurrence,) and the engine was pulled out under warranty and fitted with new rings. "Now," said the factory representative, "go out and beat the hell out of it. Take it to 7000 rpm if you like—just for short bursts. But don't baby the engine. This is a Carrera GT!"

I learned what the average Carrera owner may not know—that each engine gets a 4000 rpm bench testing that lasts several hours and a full-throttle testing that lasts several minutes, before it is ever installed in a car. Power output and all pertinent facts are noted, but an oil-burner might slip by once in a long while, and mine was one of those. Anyway, much against the grain, I complied with these new instructions and forgot the manual. At around 2000 miles, the rings seated and the "dig" came back; and believe me, the acceleration on this chariot is nothing short of brutal. Low gear (with the standard 5.17 axle ratio exported to the US) is a shade too low; but in second gear the GT—as the boys call it—leaps forward with a wild, exhilarating surge that really warms the cockles of your heart. The sports *auspuff* (exhaust system) fitted to this machine is noisy, but to any true enthusiast that noise is music of a delightful kind. As the tach needle leaps to 5200 rpm and peak torque, the cam seems to grab hold and a terrific surge of power becomes available. The savage bark of the exhaust levels off to a high-pitched snarl, and before you know it the tach is indicating 7500 rpm and you have to watch out. Over-revving the GT is the easiest thing in the world, though in an emergency it will take 8000 without the slightest fuss or complaint. (Believe it or not, a Spyder which had lost fourth gear was run around Sebring for hours at 9000 rpm without ill effects!)

The clutch on this mechanical jewel is of the racing variety, and therefore not smooth in the Porsche Super tradition. It will take a terrible beating without any sign of slipping on race-course starts, but it resents any attempts to slip it or ease it in gently while inching forward in a traffic jam. Try this and you are rewarded by a violent shudder that racks the whole transmission and rear end. It is a distinctly unpleasant sensation.

The all-synchromesh gearbox is, of course, a dream and impossible to outshift, no matter how fast you play tunes up and down through the ratios. The GT is one of the very few production sports machines with the pedals so located that you can heel-and-toe without dislocating your ankle or torturing your instep. This naturally adds to the joys of fast shifting, though with the six teeth in the differential pinion, the rear end emits a subdued high-pitched whine that was absent in former Carreras. There is also some gearbox noise which would be inaudible if the usual plastic base soundproofing undercoat had been applied.

It is hardly worth mentioning that the GT's brakes are fantastically powerful and quite fadeproof, yet do not demand abnormal pedal pressure. When you apply the brakes you stop *now*—not in a little while.

In the comfort department, the oversized interior width of the Reuter-built Porsche body is too well known to need recapitulation. The GT's bucket seats are not the lavish armchairs of the regular models, but they do a fine job of giving your back and hips support just where you need it. The rear seat has been left in on the GT, though the black leather upholstery is a token affair. Still, it will accommodate two youngsters or an enormous bag of groceries and one child, or enough suitcases for a two-week holiday.

My car has a somewhat harder ride than the average GT with the regular equipment shocks, but I purposely set the Konis to give the required stiffness for competition. And talking of that, high-speed cornering on the Carrera is beyond reproach, provided you use the tire pressures recommended for racing. With 26 lbs in the front and 27 in the rear, you can break the tail loose in the secure knowledge that the machine will respond to correction in the normal manner. There is no danger that the slide will become an uncontrollable spin, as in former years.

During the initial stages of ownership, I encountered some other (though minor) annoyances. The green oil pressure warning light went on the fritz; the idling adjustment control on the dash was not connected to the carburetors; the folding canvas top was so anchored that it was impossible to latch it to the top of the windshield—yet there is no so-called "one man" top in any sports car that is easier to put up or better looking. Literally in about five seconds, with a flick of the hand, you can flip up the Carrera Speedster's top, ready for latching.

Because of the rugged engine's almost unlimited capacity for revving, tremendously high average cruising speeds are possible, despite the low axle ratio. At 80 mph you think you are doing 50, and at 100 you might concede to about 70. With 8½ quarts of lubricating oil circulating in the dry sump system, it is almost impossible to overheat the engine. In addition, the turbo-type cooling fan in its aluminum casing circulates 39 cubic feet of air *per second!*

Detail finish on the GT, as on all Porsches, is absolutely beyond compare with that of almost any other sports car. No matter where you look, you will not find anything that is shoddy, skimpy or "marginal" in its ability to fulfill a purpose. Unfortunately, the rollbar that went with my GT was stolen somewhere in transit, but the rollbar posts welded into the chassis formed an immensely strong base for a rollbar of my own design, which is quickly and easily removed by detaching three bolts.

What *don't* I like about the GT Carrera? Well, for one thing, its inability to keep the plugs clean on anything but a long fast drive. With two plugs per cylinder, this means changing eight almost inaccessible spark plugs, countersunk into deep holes in the cylinder cooling fins. To do this, you need a special wrench (double-jointed, spring-loaded and fitted with a special suction cap), the dexterity of a trained octopus and the tenacity of a leech. Even then it takes about two hours to change all the plugs and don't lose that wrench, by the way; it's an $18 item.

If you use the car on local shopping trips for a week, or to drive to the station, it's a cinch that within a week the plugs will be totally fouled-up and need a good cleaning even if you use the standard touring jets and venturi. You will know this when the engine loses its zip and life and your forward progress becomes a sluggish affair, utterly alien to a healthy GT. It is not enough to run up to 6000 rpm for short bursts through the gears, or even 7000. The moment you get messed up by two or three traffic lights and are forced to drop to 3000 rpm, the benefits of that high-revving bout are nullified. The GT engine absolutely hates to idle, and in my experience, should be warmed-up for at least three minutes at 2000 rpm before that morning start.

Then there is the trifling matter of changing the main jets, venturi, air correction jets and mixture tubes for racing *plus* the spark plugs, of course. I learned the hard way, but the correct drill is to make all the carburction changes before you start for the race. When you arrive at the course, your touring plugs will, of course, be fouled-up by the rich mixture and you can then change to the racing plugs, warm-up at 3000 rpm and start practise. For the return trip, remove only the racing plugs, replace your touring plugs (which should be cleaned in the meantime), and take off. When you get home, you can reconvert to touring jets and venturis, and you should keep a spare set of touring plugs. (Champion N3 are suitable) as replacements. Total conversion to or from the racing kit takes about half a day of intensive labor. But the results are no doubt worth every moment of this ordeal. Do *not*, however attempt to install the racing kit and plugs an hour or so before the race, or you will never see the starting line.

The reason, by the way, why they modified the distributors drives so that they now derive motion from the crankshaft, is logical enough. When the distributors were camshaft-driven, each time a valve adjustment was made, this had a direct effect on ignition timing. And this is critical on the GT. Two degrees too much advance can burn a piston.

Another thing which I find somewhat pointless in the GT is the set of spun aluminum hubcaps supplied with this model. Hubcaps are always removed for racing, and for daily use, what do a few extra ounces matter? It would make a lot more sense to install the regular steel chromed hubcaps. The present ones dent if you so much as breathe over them.

But taken all in all, what a magnificent machine—what an enthusiast's dream is this GT Carrera! It is in a class by itself.

—**John Bentley**

Grand Touring tank is just that with 21 gallon capacity. Space left over under hood is for spare tire and tools. Bumpers—fore and aft—are easily removed for racing.

GT is easily recognized by carburetor intakes set into engine deck lid. Air is drawn in through louvers on either side of main cooling-air intake.

Porsche has tremendous acceleration for its engine size. Here, its wind-cheating nose lifts as the back wheels bite into the track at Thompson, Conn.

SPECIFICATIONS FOR MODEL

Porsche Carrera Gran Turismo Speedster

PERFORMANCE:

Acceleration thru gears

	Secs
0-30	3.8
0-40	5.5
0-50	6.9
0-60	8.7
0-70	12.2
0-80	13.8
0-100	19.9
30-50 (fourth)	7.5
Indicated maximum speed	110 mph

FACTORY SPECIFICATIONS

ENGINE:

No. of Cyls.	4
Arrangement	horizontally opposed
Valve System	double overhead cam
Bore and stroke	3.35x2.60 in.
Piston Displacement	91.4 cu. in. (1498 cc)
Comp. Ratio	9.1
Max. bhp	128 bhp @ 6400 rpm
Max. Torque	91 lb-ft. @ 5200 rpm
Carburetion	two Solex 40 PJ1 double throat, downdraft
Electrical system	six volt

TRANSMISSION:

Four-speed, all-synchromesh with integral overdrive fourth.

Overall ratios

First	15.97
Second	9.15
Third	6.35
Fourth (O-D)	4.96
Rear axle ratio	5.17

CHASSIS:

Type	Integral body and chassis
Suspension, front	Independent, torsion bar
Suspension, rear	same as front
Steering	Worm and sector (with damper)
Brake lining area	148 sq. in.

DIMENSIONS:

Wheelbase	82.7 in.
Tread, front	51.4 in.
Tread, rear	50.1 in.
Overall length	156 in.
Width	65.7 in.
Height	48.1 in.
Ground clearance	6 in.
Turning circle	36 ft.
Steering wheel turns lock to lock	2¼
Tire size	5.90 x 15
Gas tank capacity	21 gal.

REFERENCE FACTORS:

BHP per cu. in.	1.4
Weight	1848 lbs.
Lbs. per bhp	14.44
Piston speed @ peak rpm	2773 fpm.
MPH per 1000 rpm (fourth O-D)	16.07
Gas consumption	19.25 mpg.
Price	$5305

ROAD TEST

PORSCHE 1600 CARRERA DELUXE

► There is absolutely no other 1.6 liter sports car available today that has the performance and comfort of the Porsche Carrera. This figures, since with the exception of the RSK Spyder, the Carrera de luxe is the most expensive model in the Porsche line. Our test car's engine was fitted with a plain bearing crankshaft, one of the latest factory "mods" to these cars (Spyder engines still use the Hirth roller cranks). We put well over 1000 miles on it in all sorts of country, from Mediterranean sea level up to 7000-foot high Alpine passes, only reluctantly returning it to its home at the Porsche factory in Stuttgart-Zuffenhausen.

The first few miles—and thus the first impressions—were gathered on the Autobahn. A surprise was how much quieter this de luxe Carrera is to the charging GT,—for it is definitely meant to be a comfortable, safe, high-speed touring car. The thick sound-proofing material between engine compartment and occupants keeps the noise level down considerably at leisurely cruising speeds, but if one really stands on it, as the revs approach 6000 rpm the Carrera emits a solid full-chested howl. Near maximum speed, the gas pedal picks up an engine vibration that gives your right foot a gentle massage—this is one way to tell when you're flat out!

The de luxe 1600 Carrera comes factory-equipped with a pair of twin choke Solexes (40 PJJ-4) and it was our experience that although the car is not nearly as delicate in cold or hot starts, we found a definite flat spot at low revs when trying to get off the mark. Idle is steady and dead accurate even on a cold engine. After a minute or two the oil temperature rises sufficiently into the green shaded area on the dial to enable one to drive off. Thanks to two oil coolers, one up forward, just behind the front bumper, and one at the rear of the car, the last thing one has to worry about on the Carrera is oil temperature.

She runs cool all the time, even when being thrashed along flat-out hour after hour. Only after several maximum speed runs did we notice the temperature beginning to rise; also considerable oil had been "squeezed" out of the engine after running at 6800 rpm for several miles. By the end of our trip, we had added exactly three quarts of oil, apparently a good figure judging from the experience of other Carrera owners with whom we talked.

The Carrera engines are built up alongside the Spyder motors in the Porsche racing shop. Actually, there are quite a few items differing between the two types of engines. In addition to the roller crank, RS engines have different shaped pistons and combustion chambers as well as some minor items which spell the difference between the 110 hp and 135 or even 165.

The de luxe Carrera, rated at 105 hp at 6500 rpm is the "cooking" version of the twin cam dry sump racing engine; GT's come through with 110 horses. Due to the fact that all of these engines are individually assembled, each one is giving slightly different performance than the other and as we found out, there are bad and good-running Carreras, even at Zuffenhausen, and in order to make ours go the way we thought it should, we revved the dickens out of it and found it came back for more.

The test car was equipped with the latest Porsche gearbox, Type 716. There have been several alterations to the synchro mechanism so that it is smoother than ever, yet stronger. Very shortly, all Porsches will be delivered with a limited slip differential as well. The new box is standard on all Porsches except convertible "D's" with USA ratios.

Acceleration feel of the 1600 Carrera gets to be impressive only when the revs begin to climb above 4500 rpm—from 5000 to 6000 she moves out from underneath you with considerable steam; in top gear the last 500 revs—up to 7000—come hard. Below 5000 the car accelerates on a

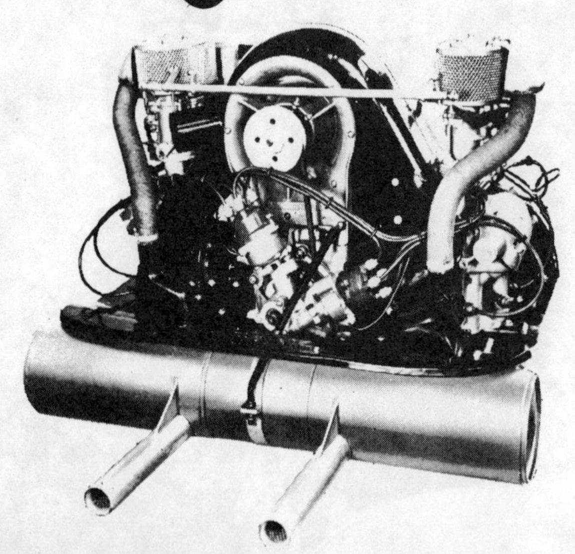

Latest 1600 Carrera engine does away with the noisy roller bearing crankshaft in favor of a plain bearing shaft. External differences are carburetor air heaters, and yet another angle for the distributors. Power? 105 genuine German horses.

par with a well-tuned 1600 Super.

The degree of built-in flexibility is another amazing thing; but something that is getting to be quite common nowadays with high performance machinery. The Carrera can be lugged down to 35 mph in top in complete safety so long as the throttle is not floor-boarded at anything below 2500 rpm. We were a bit disappointed in the maximum speed of the car, for we had been led to believe that all Carreras will top the magic 200 kph figure (124-125 miles) without any trouble. It takes considerable doing to get 125 mph out of one of the latest de luxe Carreras, and it is our opinion that early 356A's are faster than the 1959 cars.

The red area on the rev counter starts at 6000 rpm and in most of our test runs we limited the revs to 6500 through the gears, but then upon returning to the factory we were given a short ride with Huschke von Hanstein, Porsche racing manager. We complained slightly about the car's performance so all he did was take the revs up much higher than we dared. 7000 rpm in first, and in second and third the needle just went off scale, sailing up to where 8000 should be! Even at unspeakably high revs, we were impressed at the lack of that mechanical banging and chattering that one hears on pushrod Porsches when being overstretched.

Porsche interior finish has always been excellent, and that on the new Carrera Deluxe is no exception. Tach red line—at 7500—is indication of performance. Above right: Carrera engine almost impossible to work on when mounted in car.

We're not recommending Carrera owners to overrev their cars consistently but we only want to make a point that the basic Spyder engine is a mighty rugged piece of machinery and in its "cooking" form the de luxe Carrera is not quite as delicate and hard to keep running right as the earlier versions were.

The car performed extremely well both at sea level and charging up Swiss mountain passes. We did detect a slight lack of performance at 2000 meters above sea level (6500 feet) but this was hardly any surprise. On the return trip as we roared

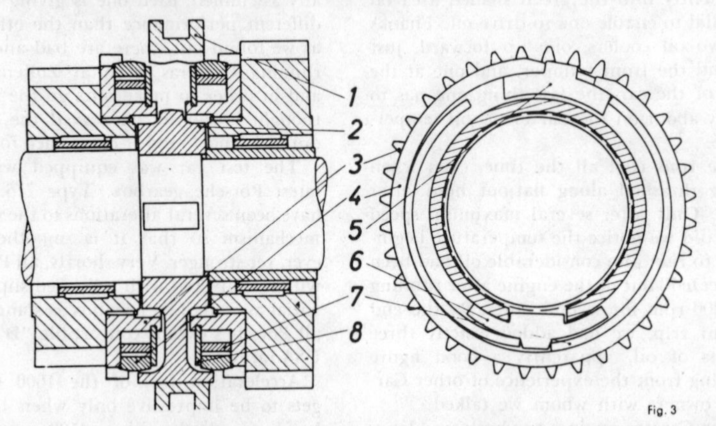

up the final 2 miles of *lacets* on the St. Gotthard at night, we noticed the green oil pressure light winking on hard left hand corners, but as soon as the car was pointed straight again, the light would go off.

The Carrera is great fun to drive in the mountains, first and second gear being well-suited for hard up-hill going. We thought first gear to be a bit too low, for in order to bridge the fairly wide gap adequately between first and second it was necessary to take low gear revs up to 6500 rpm, but after Huschke's demonstration, perhaps it doesn't matter.

Handling and road-holding of our test Carrera was markedly effected by Michelin "X" tires. The "X" is indeed a controversial tire, and let us say right now that with X's fitted to the latest 1600 Porsche 356A one goes around corners as if on a string, and this is no tired simile. The amount of understeer is formidable unless the driver dumps the car violently into a tight corner, and on severely tight hairpins we found it was actually possible to break the front end loose.

From the standpoint of wear, Michelin "X" have no equal, for where 5000 miles on a conventional tire is the absolute maximum, one can expect an easy 3000 additional hard driving miles—8000 miles on a set of tires for a Porsche is long indeed. (Ed. Note: This is under fast European conditions. If you can't afford the trip, visit Mexico.) Rear end breakaway definitely comes at a much later moment and a driver must become accustomed to this before he can drive safely on "X's" under all conditions.

We have nothing but good words to say for the ride of the Carrera; this is due largely to the Koni shock absorbers with which it was fitted. Konis are standard on all Carreras and on other Porsches on request.

Since the normal type of forced air heating does not function efficiently with the 1600 GS motor, all Carreras are equipped with a gasoline heater. The unit seems to be trouble-free, safe and exceedingly efficient, though a slight effect on fuel consumption is to be expected. The unit is fitted in the engine compartment at the extreme rear, further filling the already impressive Carrera engine room.

To work on one of these cars you need the fingers of a professional pickpocket. A 21-gallon fuel tank takes up the better part of the front compartment and leaves only space behind the seats for bags —but as experienced Porsche-pushers know, this always has been the real trunk anyway.

As in any car, the Carrera had its failings—most troublesome was its inability to get off the mark smartly without slipping the clutch. Apart from this, we enjoyed driving the car tremendously; she'll cruise all day long at 100 mph and is equally at home in the midst of city traffic. With the Carrera's flexibility in the gears, one is able to overtake or to shoot through narrow gaps almost at will—all in all a very satisfying car to drive.

—jla

TOP SPEED:
Two-way average ....118 mph

ACCELERATION:

From zero to	seconds
30 mph	3.8
40 mph	6.2
50 mph	8.2
60 mph	10.8
70 mph	14.4
80 mph	17.7
90 mph	22.8
100 mph	28.4
Standing ¼ mile	18.4
Speed at end of quarter	82 mph

SPEED RANGES IN GEARS:

I— 0- 35	(6500 rpm)
II—18- 60	(7000 rpm)
III—25- 88	(7000 rpm)
IV—35-118	(6800 rpm)

FUEL CONSUMPTION:
16½-23 mpg

POWER UNIT:
Type ....Air-cooled, flat four
Valve Operation..........Dohc, shaft driven
Bore & Stroke...........3.44x2.60 in (87.5x66 mm)
Stroke/Bore Ratio 0.75/1
Displacement..96.8 cu in (1588 cc)
Compression Ratio 9.0/1
Carburetion by....Two Solex 40 PJJ —4 twin-chokes
Max. Power....105 ps @ 6500 rpm
Max. Torque...................89 lbs-ft @ 5000 rpm
Idle Speed ....1000 rpm

DRIVE TRAIN:

Transmission ratios test car	overall ratio
I—3.09	(13.79)
II—1.94	(8.58)
III—1.35	(5.93)
IV—0.96	(4.25)

Final drive ratio—4.43 (no other available)

CHASSIS:
Frame....Welded sheet-steel punt
Wheelbase ....83 in
Tread, front and rear..51½, 50 in
Front Suspension......Ind., trailing arms, torsion bars, and anti-roll bar
Rear Suspension..........Ind., swing axles, trailing arms, torsion bars
Shock absorbers....Koni telescopic
Steering type..2F worm and sector
Steering wheel turns L to L......2½
Turning diameter, curb to curb 36 ft
Brakes........11 in bimetallic drums
Brake lining area..............122 sq in
Tire size 5.90x15

GENERAL:
Length ....156 in
Width ....65½ in
Height ....51½ in
Weight, curb (¼ tank)....2100 lbs
Weight, as tested (one up) 2240 lbs
Weight distribution, F/R as tested...........58.6/41.4
Fuel capacity ....21 U.S. gallons

RATING FACTORS:
Specific Power Output 1.09 ps/cu in
Power to Weight Ratio, as tested ....21.3 lbs/hp
Piston speed @ 60 mph 1500 ft/min
Braking Area, as tested 109 sq in/ton
Speed @ 1000 rpm in top gear ....17.3 mph

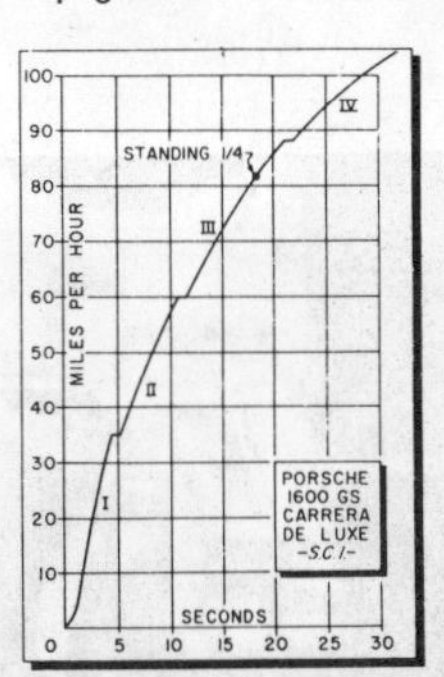

Rear view of a type 1600 engine ready for installation, showing the baffling around the cylinders to give close control of the path of cooling air

PORSCHE

ALTHOUGH the history of the automobile has now spanned more than 70 years, there have been surprisingly few great names among the designers; surely among the greatest must rank Dr. Ferdinand Porsche. In many ways he can be compared with Sir Henry Royce, for each came from a humble home, with little or no technical heritage to push him into the field which he subsequently graced for so long. Both were intuitive engineers, commencing their careers in the electrical field, without the advantages of degrees and formal schooling; Dr. Porsche's title was an honorary one bestowed by the Austrian government.

But whereas Royce was basically a fine development engineer who perfected other people's ideas by the application of sound and well-executed engineering principles, Dr. Porsche was a creative artist who tackled new problems with clear logic, and was never afraid to apply hitherto untried ideas. In this respect his achievements are more comparable with those of Dr. Fred Lanchester.

From the Austro-Daimler works, where he was head of the development section at the age of 30, Dr. Porsche moved to Daimler, in Germany, before their association with Benz, back to Steyr in Austria, and finally formed his own design office. Here he evolved the designs for the fabulous Auto Union racing cars, the Volkswagen, the Cisitalia Grand Prix car, and, finally, the car which still carries his name. In these latter two projects Dr. Porsche was assisted by his son Ferry, who has inherited many of his father's characteristics and abilities, and Julius Rabe, who joined him at Austro-Daimler in 1913 and is today chief engineer at Porsche.

Like Henry Ford, Dr. Porsche, who died in February, 1951, was a visionary, and clearly foresaw the day when every household would possess at least one car. Indeed, he prepared the design of a *Volkswagen* or People's Car two years before the Hitler government made it a top priority project. Initial attempts to manufacture this car were made in 1931 at Zundapp, and later at NSU. All the initial developments of the four-cylinder, horizontally-opposed air-cooled engine, swing axle and torsion bar suspension were carried substantially unchanged into the post-war Porsche by way of the final Volkswagen conception. All Porsche projects are given a project number, that for the Volkswagen, for instance, being No. 60. Thus, the Porsche had the number 356 and the Spyder 550—the basic type numbers by which they are still known today.

It was during the time that Dr. Ferdinand Porsche was held in custody by the French government that his son Ferry conceived and built the pilot model of the present range. It was evolved from basic VW parts and the engine was placed forward of the rear wheels, an arrangement subsequently abandoned for the production saloons but again taken up when the type 550 Spyder was introduced. It says much for the affinity of engineering thought between father and son that Dr. Porsche remarked, when seeing the vehicle for the first time, that he would have changed little had he been responsible for the entire design from its inception. This first vehicle used the 1,131 c.c. Volkswagen engine, which had been increased in power to 40 b.h.p. from its original 25 b.h.p.

When Porsches were first put into series production, the engine capacity was 1,086 c.c. This was achieved by retaining the VW stroke of 64mm (2·52in) and reducing the bore to 73·5mm (2·89in). In these early days the majority of the engine components comprised standard VW parts, the important exception being the cylinder head. Gradually, as development proceeded and production increased, Porsche redesigned and manufactured their own components in the 356 push-rod series. A standard VW crankcase was used until the end of 1954, but Porsche are no longer dependent upon the Wolfsburg concern for the supply of parts, although, in fact, many are still interchangeable. Thus, the development of the Porsche engines falls into two distinct phases, the push-rod type 356 evolved from the basic VW, and the bevel-driven overhead camshaft unit designed in 1952, which forms the basis for all Spyder, Speedster, Carrera and racing units. Throughout the development of the type 356, from the original 1,086 c.c. unit to the 1,582 c.c. version, the same cylinder centres of 102mm (4·0in) on each side and a transverse offset of 37mm (1·46in) between opposing cylinders on adjacent crankpins have been maintained.

In all engines there are certain design features which control the overall proportions, and in any horizontally opposed cylinder layout it is the crankweb thicknesses and bearing widths for the big ends and centre main which are the important design parameters. It is a popular misconception, particularly when fins are required for an air-cooled unit, that the spacing of the cylinders is the basic determining factor.

During the develop-

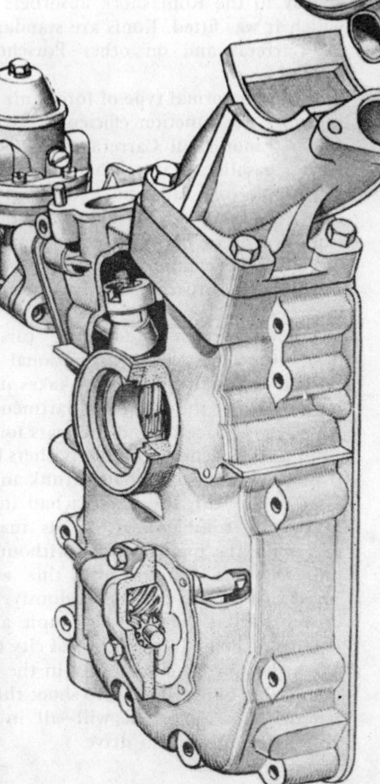

A half crankcase of the overhead camshaft type 550 engine with the crankshaft in position. In this version the crank webs are thicker than for the type 356 and there are four counterweights. The roller bearing crankshaft version of this engine is now used for racing only. On the end of the half-speed shaft there are two back-to-back bevel gears, the first stage in the shaft drive to the camshafts

The development of the type 356 push rod, and type 550 o.h.c. horizontally opposed four-cylinder engines

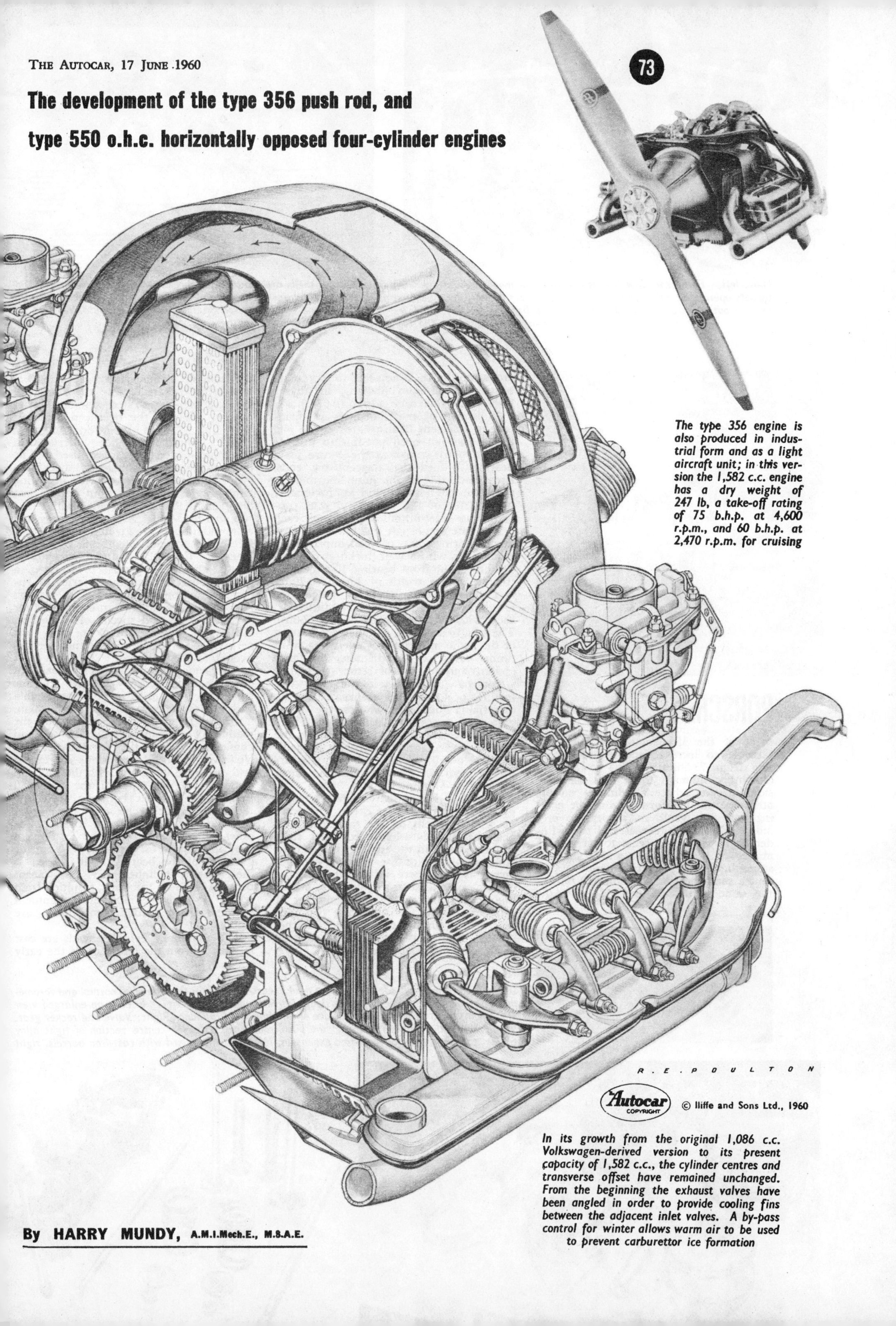

The type 356 engine is also produced in industrial form and as a light aircraft unit; in this version the 1,582 c.c. engine has a dry weight of 247 lb, a take-off rating of 75 b.h.p. at 4,600 r.p.m., and 60 b.h.p. at 2,470 r.p.m. for cruising

In its growth from the original 1,086 c.c. Volkswagen-derived version to its present capacity of 1,582 c.c., the cylinder centres and transverse offset have remained unchanged. From the beginning the exhaust valves have been angled in order to provide cooling fins between the adjacent inlet valves. A by-pass control for winter allows warm air to be used to prevent carburettor ice formation

By **HARRY MUNDY**, A.M.I.Mech.E., M.S.A.E.

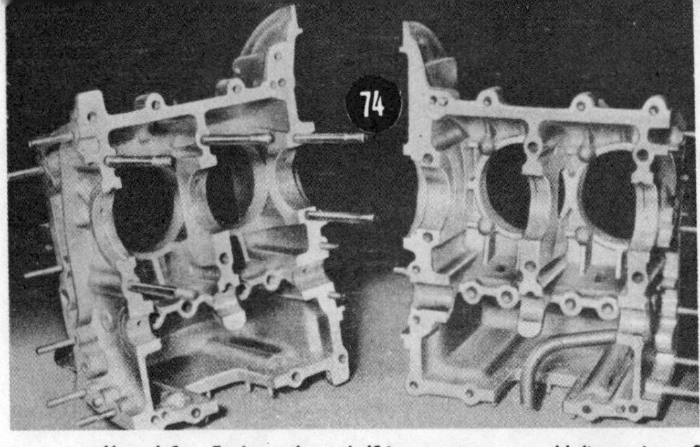

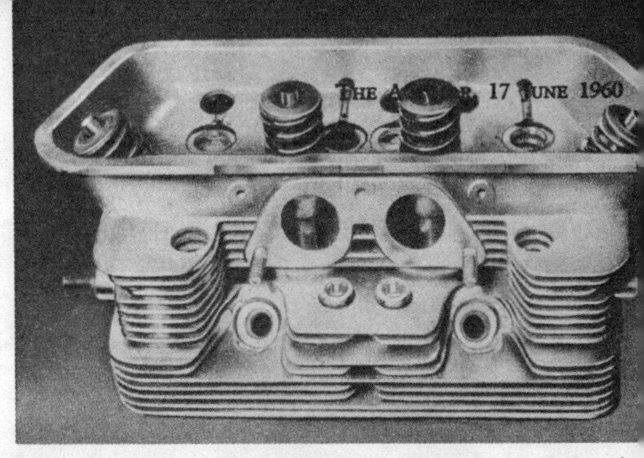

Above left: Each crankcase half is a permanent mould die-casting. Separate bearing shells are used for the three mains, but the camshaft and tappets operate directly in the casting. Above right: A fully machined and assembled cylinder head showing the individual inlet ports with the exhaust outlets at each end; the valves are in line transversely. Below: Crankshaft and connecting rod assembly. On the type 356 engine the stud extensions—nutted from the top—are formed integrally with the caps

PORSCHE . . .

ment of the 356 engines it has been possible to increase the cylinder bore size from 73·5mm (2·89in) to 82·5mm (3·25in). Although there is a flattened side between adjacent cylinders on the otherwise circular formation of the cooling fins, this would not appear to prejudice still further increases of the cylinder bore. It would seem to be more a matter of bearing loads and crankshaft strength, for the effective width of the centre main bearing is 20mm (0·79in)—the same as that of the big ends; the crankweb thickness adjacent to each main bearing is 11·5mm (0·49in), the flying webs being thicker at 17mm (0·67in).

Indicative of the difficulties in maintaining sufficient bearing area is the fact that it has not been possible to provide a continuous groove in the centre main bearing to feed the adjacent big ends. From each side of the joint line there is a groove extending round approximately one-quarter of the periphery, thus providing an intermittent feed during each half of one crankshaft revolution. These main bearings are of lead-impregnated aluminium, with a wall thickness of 5mm (0·20in). The front bearing, like the centre one, has a width of 20mm (0·79in), and the rear one, which is flanged for location purposes, is 29mm (1·14in) wide, each with a diameter of 50mm (2·0in). There is, in addition, an outrigger bearing of the same material located in the front timing cover, with a diameter of 40mm (1·58in) and a width of 18mm (0·71in).

Porsche engines retain the basic VW features of a two-piece crankcase, but the material is aluminium, whereas the Volkswagen now uses magnesium alloy. There are half-bearing housings in each for the crankshaft and for the camshaft immediately below it. The camshaft is gear-driven from the front end of the crank by steel-to-aluminium alloy gears, the front end of the camshaft operating the oil pump through a tongue drive. There are no separate shells for the three camshaft bearings, the case-hardened steel shaft running direct in the casting. One economic advantage of the opposed-cylinder layout is that there is half the number of cams on the shaft required by an in-line engine having the same number of cylinders—each cam operating two tappets, which, in the case of the Porsche, are of true mushroom type, working directly in the crankcase casting.

Originally, all Porsche engines had cast-iron cylinder barrels, and these are still used in the lower-rated standard series engine. For all the overhead camshaft and Super series push-rod range of engines, aluminium alloy barrels with chromium-plated bores are used. The three advantages claimed are longer bore life, lighter weight, and lower operating temperatures resulting from better heat dissipation. These light alloy barrels are supplied by Mahle. After fine boring, the aluminium bores are given a dimpled surface of closely controlled pitch and depth. They are then plated and finally honed so that the finished surface is one of a smooth bore, broken by a series of enlarged "pinholes" which act as oil pockets. Extra running-in is required for these cylinders and special oil is used at the works during break-in and test periods, and is retained in the engine as delivered to the customer. After the first recommended change a normal detergent oil is used, at which stage the oil consumption is claimed to be comparable with that of the cast-iron cylinder version.

On the type 356 range of engines, the method of attaching the big end cap and connecting rod is the same as on the Volkswagen. Stud extensions are forged and machined integrally with the cap, and nutted from the top; for the type 550 range there are set bolts threaded into the cap. In each type, one of the studs is a close fit and used for location, the other one having clearance. To prevent the cap twisting on assembly, the outside diameter of the bearing shell is used as the location member. By this means bending loads on the bolts, resulting from trying to achieve absolute alignment between three fitting components, are eliminated.

The aluminium cylinder heads are cast in pairs. From the inception of the early

Left: Two types of cylinder barrels are used, a cast-iron one for the type 1600 normal and chrome-plated aluminium for the Super versions, Carrera and Spyder series. Inset is an enlarged view of the honeycomb chrome-plated surface of an aluminium cylinder. Right: Valve and rocker gear. When light alloy barrels are used the push rods have a longer centre section of light alloy tubing, to compensate for increased expansion, left, than those used with cast-iron barrels, right

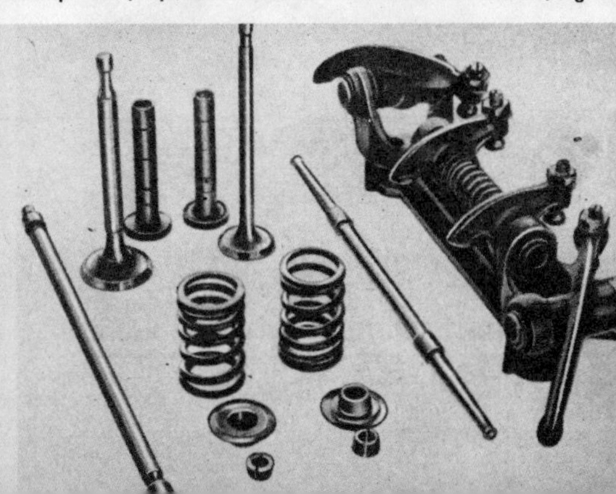

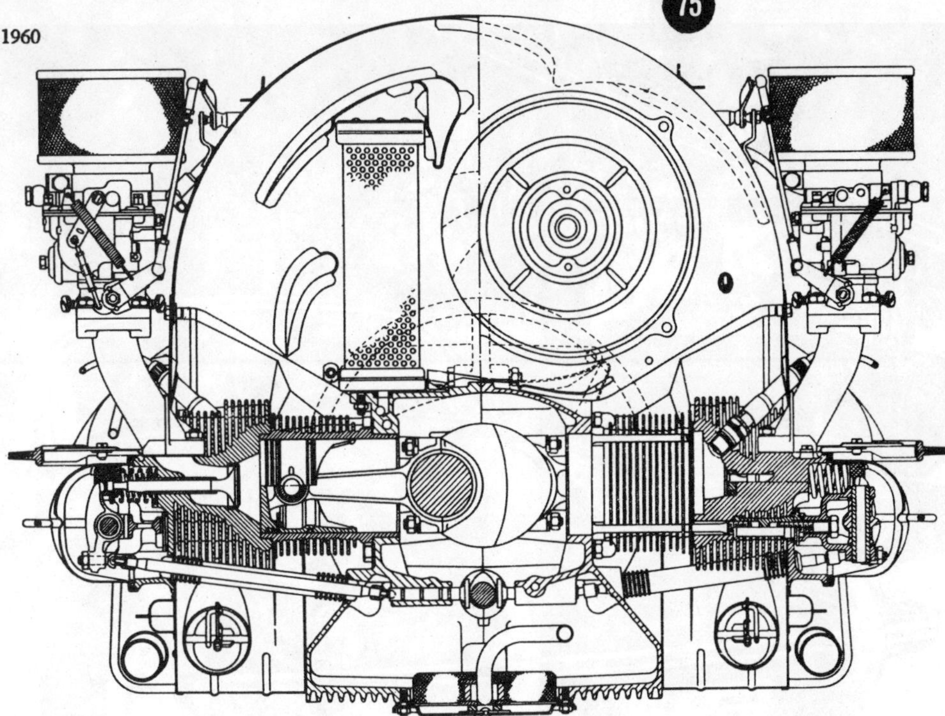

Cross-section of a type 356 engine. Valves are operated from a low-mounted, gear-driven camshaft with mushroom type tappets. The vertical honeycomb stack adjacent to the cooling fan is an oil cooler, used on all versions of the engine

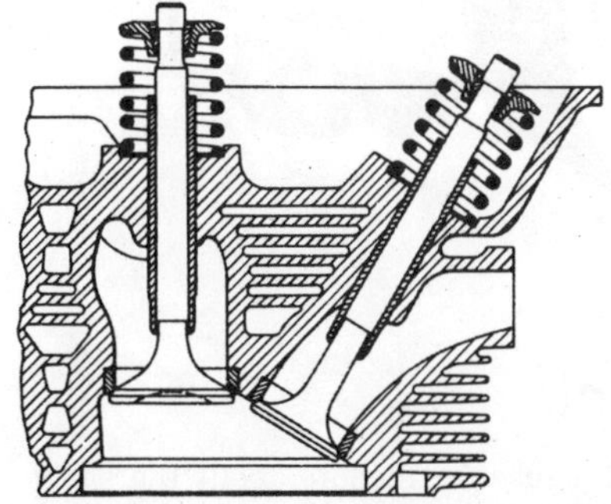

Below: Cross-section through the valves and combustion chamber. The bronze guides are retained by interference fit, achieved by heating the cylinder head at the same time that the seats are inserted. Below right: Net horse-power ratings, in accordance with continental DIN standards, for the main basic types of engine

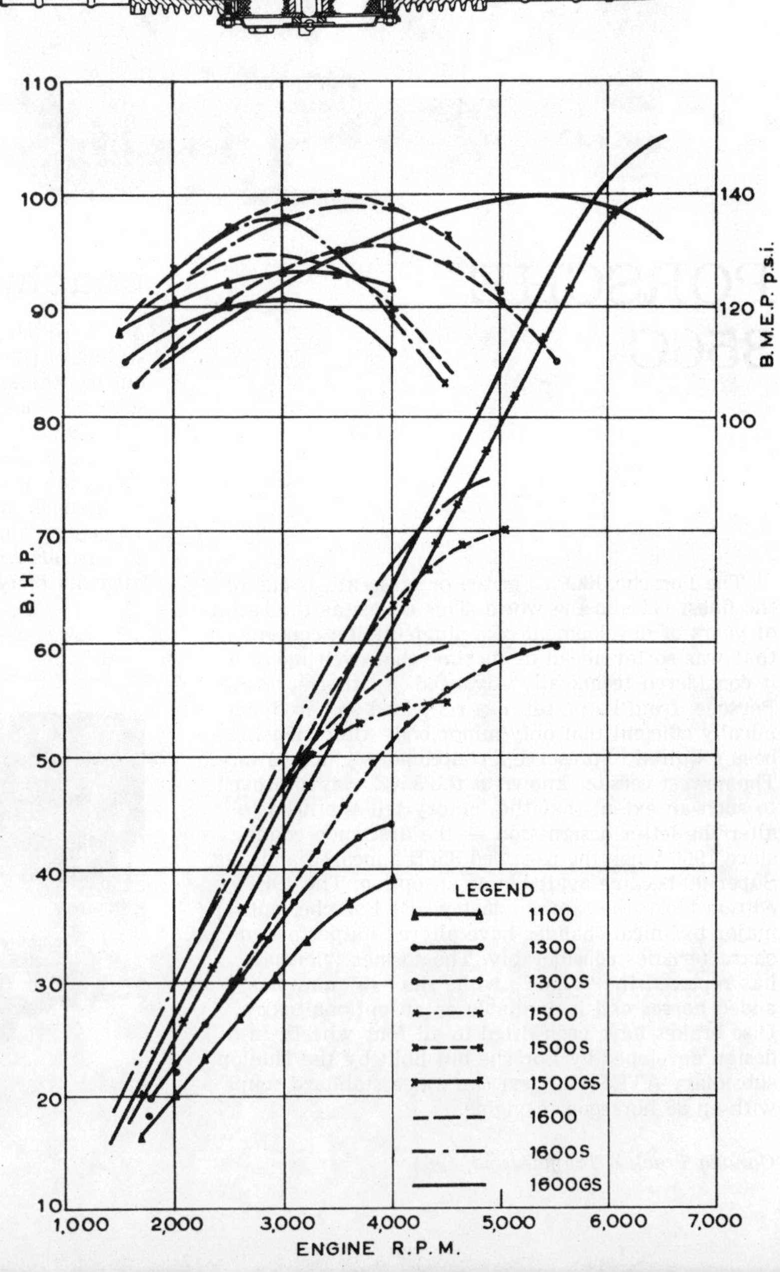

VW derivation, Porsche have used their own valve and porting arrangement, with the two separate inlet ports on the uppermost face and exhaust outlets at each end. The four valves in each head are in line on the longitudinal axis, but when viewed from the top the exhausts are angled at 30deg to their respective inlets. This was done to provide cooling passages between them, which would have been impossible, or at least ineffective, if the exhausts were placed vertically beside the inlets.

Cooling air is provided from a shrouded impeller-type fan, belt-driven from the crankshaft at 2·65 times engine speed, and mounted on the end of the dynamo. For all push-rod versions, the 230mm (8·78in) fan has a single entry, but for the high performance sports and racing versions of the twin-camshaft units a 240mm (8·82in) double-entry type is used. For the single-entry type the air flow is 10 cu. ft. per sec. at 2,000 r.p.m., and 20 cu. ft. per sec. at 4,000 r.p.m. The power required to drive the fan is 4·9 h.p. at 4,500 r.p.m., 6·6 h.p. at 5,000 r.p.m., and 9·0 h.p. at 5,500 r.p.m.

Almost enclosing the engine is a sheet-metal cowling containing carefully sited directional baffles to route and grade air flow across the cylinders. Within this cowling also there is an oil cooler on all engines, the system incorporating a by-pass which brings the cooler into circuit at a predetermined temperature.

Technical features and development of the bevel-driven overhead camshaft engine series will be analysed next week.

PORSCHE 356C

☐ The Porsche, like a Ferrari or Maserati, is one of the finest GT's in the world. This comes as the result of years of development of a single-design concept that was so far ahead of its time that even today it it considered technically advanced. Furthermore the Porsche, from its outset, has remained so aerodynamically efficient that only minor body alterations have been required to preserve a contemporary appearance. The newest version, known as the 356C, was improved to such an extent that the factory felt the need to alter the letter designation — the first such change since 1960 when the re-styled 356B appeared and the Super 90 became available as an option. The 1964 edition looks the same as last year's Porsche, but major technical changes have altered its performance characteristics considerably. The former "normal" has replaced the "super", while the "90" now has added horses and is available as an optional extra. Disc brakes have been fitted to all four wheels, in a design developed by Porsche but built by the Dunlop subsidiary ATES. Our test car was a standard coupe with an 88 horsepower engine.

coachwork

There has been cause for dispute about the early styling of the Porsche. To some it was pure and pretty, but to others, particularly those who like fins and squared lines, the Porsche resembled a turtle with wheels. The higher fenders and stylized bumpers added in 1960 improved the looks in many eyes, while enlargement of the rear window, with twin air grilles at the back, has changed the appearance even more. The current model, in our opinion, is a welcome curvaceous relief in these days when most bodies are basically box-like. *(continued next page)*

PORSCHE 356C

interior

It is often said that Porsche appreciation begins as soon as one slips inside. This is because Porsche's interior is almost unmatched for comfort and quality. It begins with the seats — built by Reuters, famous for the luxury of their bucket-type, fully-reclining chairs. These are scientifically designed to provide the right amount of support exactly where needed. Gentle contours provide lateral grip without being too extreme for easy access. In the rear, a pair of small bucket seats are excellent for children, but adult use is restricted to short hauls. The backs may be folded down into a spacious platform. Adjustability of the seats permits tallest riders to stretch at length and there's ample elbow room and head space. The pedals are slightly offset. Vinyl is used in most of the interior except for some cloth trim on the seats, while carpeting lines the foot wells. The rubber matting on the floor is more practical than appealing. Large map pockets are fitted on the doors.

instruments

Porsche instrumentation is not impressive for those accustomed to the sight of dials and switches in every corner. There are merely three circular instrument faces, and the knobs are simple black push-pull type. But the important fact is that everything seems to be in the right place. All the instruments are easily read . . . all the switches are easily reached. And all the required information is available: speedometer, tach, resettable trip odometer, oil temperature, fuel. The steering wheel is of masterful detail design and the gear lever is within easy reach and has a good feel in the hand.

engine

The Porsche engine is familiar to most enthusiasts. It was inspired by the Volkswagen, which was designed by the same technical genius, although it shares no parts. The layout is flat four, opposed, overhead valve, over-square, air-cooled, with twin Solex carbs, very light and very strong. The latest version is similar to the former super, but has been developed to provide greater torque in the low rev range without any of the super's fussiness. There is no choke, cold starting being accomplished by pumping the accelerator five times before operating the starter. A hand throttle is fitted to keep the revs high when idling with a cold engine.

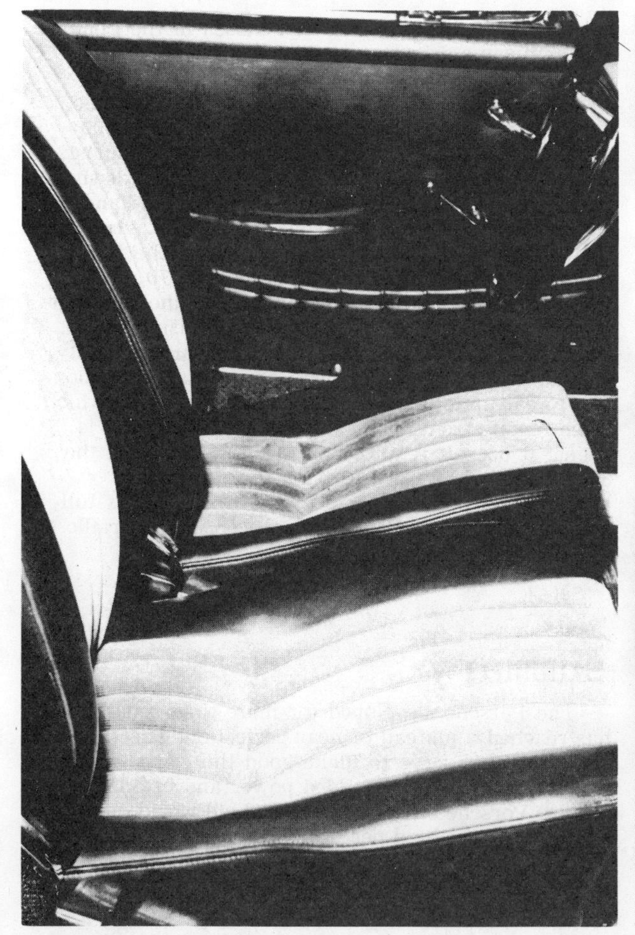

trunk

Luggage space is undoubtedly a weak point with Porsche. Admittedly the trunk has been enlarged slightly in recent years and one may carry an enormous number of suitcases behind the seats or add a luggage rack. But with the back seat occupied there is very little space for luggage or groceries. Porsche owners must simply learn to adapt. One major improvement has been made though: the gas filler has been moved from inside the front hood to a fender location where it is conveniently opened by a latch within the car. An attached cloth keeps careless pump men from splashing fuel on the glamorous fender.

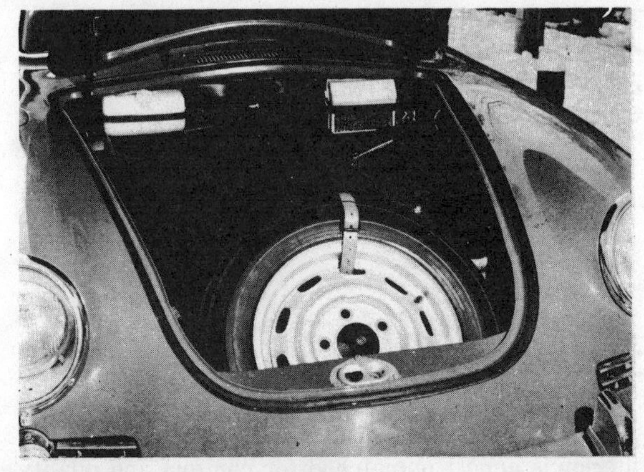

handling

Porsche-oriented drivers are in for a delightful surprise when they climb behind the wheel of a 356C. Roadability is amazing for this rear engined car and oversteer is gone . . . at least to the point of really fast cornering. Until extreme cornering speeds are reached, the Porsche handling remains neutral. Combined with the lightness of the steering it makes cornering a delight, even for the novice. The transition to oversteer is gradual, and when reached permits the car to be drifted in classic fashion. The disc brakes are among the best we have tested. The ride is smooth, thanks to the all-independent torsion bar suspension, and this also contributes to stability when cornering on bumpy surfaces. Many enthusiasts use Porsche for reliable competition motoring, and the swift handling is one of the main reasons for its popularity. The combination of qualities which add up to superb control on racing circuits or for rough road driving is noticeable, even during a short testing period.

performance

There's a world of difference awaiting the owner of an older Porsche when he tries the new one. The acceleration is greatly improved all the way through the speed range. Tle low-down torque gives excellent acceleration in the 0-60 area, while the added power keeps the movement rapid up to the 80 mark and more, before the speed curve begins to taper off noticeably. Gear ratios are well-chosen in the Porsche although some drivers may wish for a lower fourth in order to maintain the acceleration rate at high speeds. The present gearing gives an overdrive effect, keeping the revs low at high speed cruising (4,000 at 90 mph). The Porsche gear box operates almost effortlessly and with no sponginess. Movement is short, yet always into the desired cog. Fast as the hand can move, it's nearly impossible to beat the synchro. In fact it's so efficient (and strong) that the tester once put his own Porsche into reverse at 70 mph while racing with the result that the engine blew but the transmission didn't! High speed cruising in comfort is still the Porsche's most endearing quality. And this is where the car will appeal to those who like to eat up long mileage without the fatigue which often results from distance driving. The Porsche is eager; it doesn't require pushing. But don't sell the 356C model short for all-round use. Its nimble handling and flexible performance range allows full-time appreciation of the car . . . even in city traffic.

summary

The Porsche, developed for more than a decade, has reached a plateau of near perfection. The 1964 improvements serve to make good things better, in terms of performance, added power and braking ability. Porsche offers comfort and old-world quality along with race-bred performance and handling. It is a joy to drive over any terrain. It's limits are largely those of the driver.

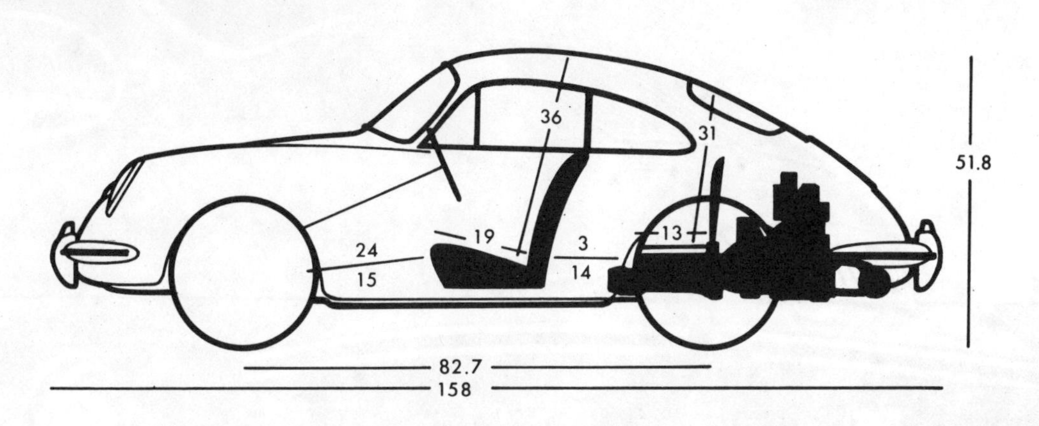

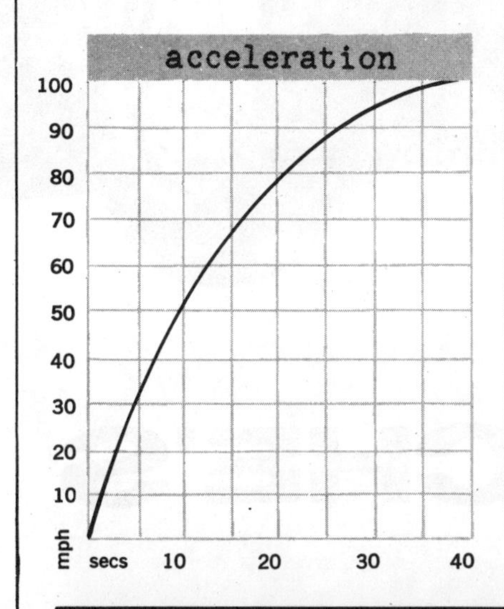

performance

ACCELERATION
0-30— 4.2 seconds
0-40— 6.6 seconds
0-50— 8.9 seconds
0-60—12.5 seconds
0-70—15.9 seconds
0-80—21.8 seconds

SPEEDS IN GEARS
1st—29
2nd—52
3rd—78
4th—112

PASSING SPEEDS
30-50— 5.9 seconds
40-60— 6.4 seconds
50-70— 9.0 seconds
60-80—12.6 seconds

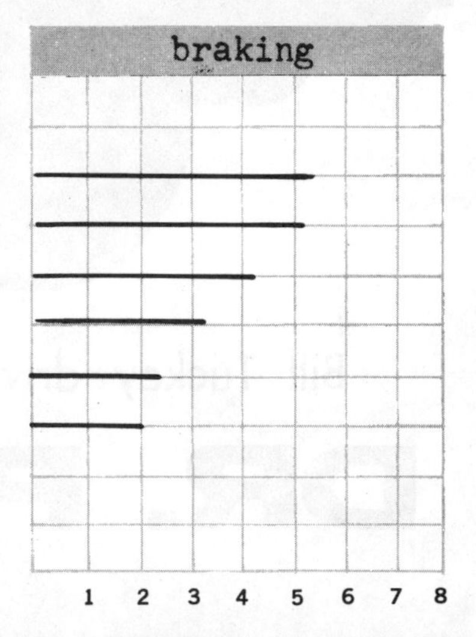

PORSCHE 356C

TEST CAR COURTESY EGLINTON CALEDONIA MOTORS LTD

ENGINE—

Location:	rear.
No. of Cylinders:	4.
Head Type:	OHV.
Compression Ratio:	8.5:1.
Carburetors:	two dual downdraft, Zenith.
Cooling:	air.
Bore:	3.25"
Stroke:	2.92"
Displacement:	1,582 cc (96.5 cu. in.)
BHP:	88 SAE @ 5,200 rpm.
Torque:	90.5 lb. ft. @ 3,600 rpm.

TRANSMISSION—

No. Forward Speeds:	4.
No. Synchro:	4.
Gear Ratios:	1st: 3.09:1; 2nd: 1.765:1; 3rd: 1.13:1; 4th: 0.852:1.
Axle Ratio:	4.428:1

DIMENSIONS—

Wheelbase:	82.7"
Track f and r:	51.4/50.1.
Length:	158"
Width:	65.8"
Height:	51.8"
Ground Clearance:	5.9"
Fuel Capacity:	10.5 imperial gallons.
Weight, Curb:	2,060 pounds.
Weight Distribution	
front:	44%.
rear:	56%.
Tire Size:	5.60x15.

STEERING—

Type:	worm gear.
Turns Lock to Lock:	2.5.
Turning Circle:	33.5 feet.

SUSPENSION—

Front:	trailing arm type, torsion bars, anti-roll bar.
Rear:	swinging half axles, torsion bar.

BRAKES—

Front:	disc.
Rear:	disc (plus drum for hand brake).

PRICE AS TESTED— $5,295.

Bill Tuckey drives....

DR. PORSCHE'S

FOR the benefit of the unlikely few who have never read "Kings Of The Road", or "All But My Life", Ken W. Purdy is an outstanding automotive journalist. More a superb writer than a technical analyst, Mr Purdy is also a lifelong devotee of the Porsche. Everywhere we go we find Purdy's rich prose leaping out at us. This, in the gospel-according-to-Purdy, is the best car in the world.

Up to a few weeks ago our favorite Porsche quote was that of Dan Gurney who said, on sampling the first FI Porsche: "It goes pretty well for a Volkswagen." Now all is changed. After a week with a beautiful flame-red 356C coupe from NSW Porsche distributor Alec Mildren we are inclined to think that Ken Purdy may be a master of understatement.

The Porsche magic has rubbed off on the staff of SPORTS CAR WORLD. This is a marvellous motor car; agile, lithe, responsive yet unforgiving, the Porsche is like no other car on the road —except, perhaps a Volkswagen, and then only in noise.

The Porsche tradition stretches back even further than that of its poor relation, for Dr Porsche was designing interesting automobiles in the 1920s. He was involved briefly with many organisations, including NSU and Auto-Union, but today's Porsche stands as a monument to his genius for integrated design. The Porsche has the best gearbox, the best finish, one of the most efficient bodies and the lowest wear index of any sports car in the world, regardless of price.

The C series Porsche was the first in the line to be equipped with disc brakes. This was probably because Teutonic thoroughness would not entertain disc brakes without an efficient handbrake; so they built a small drum brake into the rear discs to do the job. The bread-and-butter Porsche is available as the 356C, or the 356 SC — rare in Australia — which is uprated in compression ratio, bhp and torque output, final gearing, and tyre size, It is also considerably faster, at 115 mph against 109 mph. It is interesting to note that the factory handbook specifies 109 mph as the C's top speed, for that is exactly what we obtained.

Made in welded unit on platform construction, the 356 series follows Porsche tradition in slinging the flat-four engine behind the line of the rear axles, with the drive going forward to the gearbox. The classic engine is well oversquare, developing peak bhp at a relatively low figure of 5200 rpm but peak torque at 3600 rpm, which puts the performance area for top gear in a most useful zone of 60-100 mph. This 1600C engine still comprises the bulk of Porsche production, despite the Carrera 2-litre and the 911 six-cylinder, and retains the basic treatment of overhead valves controlled by a crankcase-mounted camshaft. However, with the 356 series came a new cylinder head and remade induction and exhaust manifolding. Although air-cooled — always a bulkier method of cooling — the engine takes up little space in the rear compartment.

The suspension has changed very little for this series: dual trailing arms and square-section transverse torsion bars at the front, plus an anti-roll bar; swing axles and torsion bars at the rear. The problem of controlling camber

SPORTWAGEN

changes through rear swing axles has been tackled by carefully assessing damper angles and by anchoring the outer ends of the torsion bars in the vertical plane.

Thus, mechanically, the Porsche has changed little. Its basic body outline has remained unaltered from the original classic styling which enables even the least alert to pick a Porsche at long range. For the 356 series the glass area was made larger, rear air intake redesigned, overriders and lighting treatment modified slightly, and a fraction more brightwork added. Not much, but it improves on something that was very difficult to improve upon in the first place.

What impressed us most about the test car was the paintwork. It seemed about 50 coats deep; lustrous, deep red, with not a single blemish even under rolled flanges and door edges. The brightwork is of the same quality, strong and hard. Inside, the red paintwork is carried through on to the dashboard, and here the taste is not quite as good. What one ends up with is an expanse of red-painted metal, no matter how you cook it, and some of our drivers felt it lacked dignity.

The interior also carries touches of Volkswagen, mainly in the coarse-weave but highly durable floor covering. We felt this could have had a more luxurious look for the price. The seats are covered in air-permeable pvc of good quality. They are well-shaped, with a definite recess for the backside and good support under the thighs; and they are wider than one finds normally. The backrests are adjustable, but not fully reclinable; and, as in most Continental cars, the adjustment for coarse and fine is a little too coarse and slightly too fine. I could not quite get the right arm distance without having to fetch a shoulder off the backrest to get third gear.

However, one could only class the driving

Driver-designed, quality controlled, the cockpit is a masterpiece of careful layout. Wheel/gearlever relationship is excellent and finish is just superb.

position as excellent, as the near-vertical wheel is well related to gearlever and leg angles. The floor is flat, except for a shallow tunnel, with deep foot wells formed by high side platform frame members, and a steeply sloping toeboard. The wheel arches do intrude into the compartment. A stiff pocket is formed in each arch curve, and the flat ledges formed by the sills make handy places on which to rest gloves, sunglasses or cigarettes. In each door is a tensioned map pocket.

The floor covering is in good rubber. The grey interior trim is carried over the dashboard cowl to stop reflections, while the sun-visors are padded and the rear vision mirror dimmable against glare.

Neatly done in black and chrome, with the Porsche emblem in the boss, the steering wheel is slightly dished, with a full horn ring. The gear lever, cranked midway, carries a heavy black knob. The window winders are terribly low-geared at 4 2/3 turns, while a stalk on the right side of the column operates trafficators and headlight flasher (pull to flash, flick to dip). It is in exactly the right place.

There is a little too much pizzazz on the dashboard for our tastes. The instruments are lettered in green, with the tachometer getting multi-stripes that eventually become red in the danger area at 5500 rpm. There are three main dials, the right one carrying fuel and oil temperature gauges, the centre the tachometer, and the left-hand one the speedometer. In the centre is the clock, topped by the fresh-air levers, and at left a lockable glovebox with a magnetic latch and an interior that slopes downward to stop loose objects leaping out.

The ignition/starter switch is at far right, but controls for heater, cigarette lighter, head and parking lights and the wiper/washer system are grouped in the centre. Everything is properly placed and within easy reach. The choke is automatic, and the passenger has a chromed grab handle for moral support. Naturally, the wipers —which cover the right area — are rheostated, as is the instrument lighting (over a very wide range), there is a map light set into the centre of the facia and coathooks in the rear. Warning lights tell when the handbrake is on, various lights are working, and if the oil pressure is failing.

The rear compartment is fascinating, because

An exhaust pipe emerges through each overrider (is this perhaps for airflow still?). Reversing lamp sits next to left rear overrider, is well-protected.

although too small for a medium-sized adult, it offers plush accommodation for smaller types. Two firm seats are formed each side of a big centre tunnel, and the backrests and cushions curve right around the body. However, the split backrests clip down to form a luggage floor, again carpeted in the heavy weave. With this and the room for soft bags at the front, the Porsche will carry a surprising amount of gear.

This is a completely integrated machine, in which everything is there for a purpose and mates to all the other functions. The Porsche rides very well over all surfaces with no suspension noise worth mentioning. In fact, the only real noise inside the car is a low level of engine buzz and a vacuum drumming sensation that develops in the rear compartment when the wind is blowing a certain way and the rear vents are closed.

This drumming is a common VW phenomenon, and can be attributed to a turbulence, due to good air flow, occurring around the driver's window and perhaps merging in a Doppler effect with the engine resonance to produce what at times was a most annoying throbbing of the air

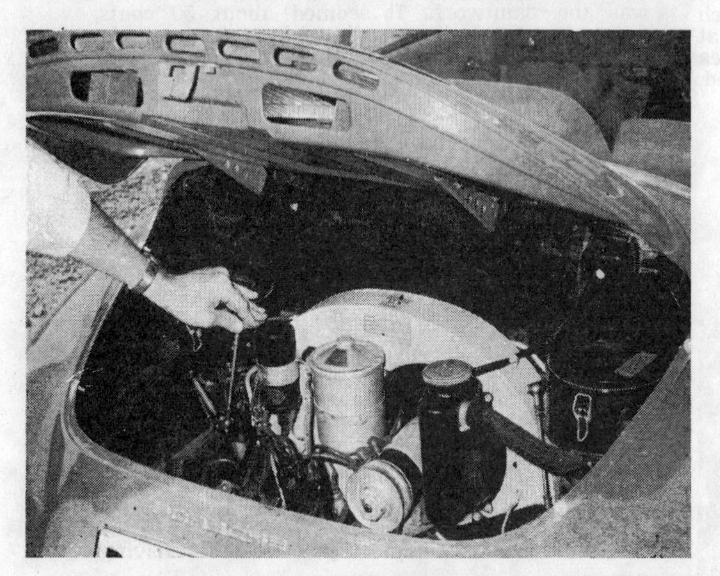

Innocuous, only half-accessible, yet very well designed and lowly stressed, the 1600cc flat four develops 75 bhp (DIN). Bonnet lock is in interior.

SPECIFICATIONS

CHASSIS AND BODY DIMENSIONS:

Wheelbase	6 ft 10¾ in.
Track, front	4 ft 3½ in.
Track, rear	4 ft 2 1/10 in.
Ground clearance	5 in.
Turning circle	33 ft 6 in.
Turns, lock to lock	3 ft
Overall length	13 ft 2 in.
Overall width	5 ft. 5¾ in.
Overall height	4 ft 3¾ in.

CHASSIS:

Steering type	2F worm and peg
Brake type	discs, all wheels, 9 in., front, 9½ in. rear
Suspension, front	independent parallel trailing arms, laminated torsion bars, anti-roll bar
Suspension, rear	independent, swing axles, torsion bars
Shock absorbers	Koni telescopic
Tyre size	5.60 by 15
Weight	18½ cwt.
Fuel tank capacity	11 gals
Approx. cruising range	330 miles

Underbonnet space houses spare, 11 gallon fuel tank, washer container and room for soft bags. Lid is counterbalanced very cleverly, opens from inside.

inside the car.

The steering is pin-sharp and quite light except at slow parking speeds, while the gearbox is lightning-fast and phenomenally good. The steering has the necessarily high castor loading to minimise the oversteering feel, but is really superb.

With a gearbox like this it is never difficult to be in the right gear at the right time. I felt first gear was a fraction too low, but the ratios are otherwise well-spaced. The car cannot be run comfortably below 30 mph in top gear, but who wants to? The standard VW/Porsche technique of snatching third as soon as a sharpish corner looms up becomes a real joy. The brakes are first-class, showing no signs of locking or fade at any time.

Possibly the most controversial part of a Porsche is its handling. It is not a car that you can drop your seat into and drive hard instantly, as you can with a Sprite, for instance. The test car was fitted with German Dunlop sports tyres, and, as expected, proved sensitive to pressures. It was a consistent oversteerer; this at first gives most drivers a little trouble, but it becomes fairly easy to sense when a Porsche is going to lose adhesion at the rear because the steering seems to move perhaps half an inch into oversteer a moment before anything happens.

Once you learn the car, a new world opens up. I remember best a series of climbing, tight corners that one normally covers at 40 mph in a family sedan; on a wet, drizzly, greasy day the 356C rocketed through them at twice that in a series of corrected slides. It is one of the few cars in which you can develop a rhythm of oversteer much like that seen when a good racing driver is hard at work with a car that is a little tail happy. The tail moves out, you put on correction, and the car balances out there for a moment until the correction takes effect; as it comes back, so you take back the lock in one smooth motion, so the car and the wheel both arrive back together. And then the same for the next corner. It becomes a sort of smooth, flowing process that is most satisfying.

From all this you may gather that we like Porsches. We do. Not as fast as some, not as spectacular as others, it does its job better and more efficiently than any other sports car in the world. Just that. #

ENGINE:

Cylinders	four, horizontally opposed
Bore and stroke	82.5 mm by 74.0 mm
Cubic capacity	1582 cc
Compression ratio	8.5 to 1
Fuel requirement	95 octane
Valves	pushrod overhead
Maximum power	75 bhp (din) at 5200 rpm
Maximum torque	90.5 ft/lbs (din) at 3600 rpm

TRANSMISSION:

Overall ratios—	
First (synchro)	13.69
Second (synchro)	7.82
Third (synchro)	5.01
Fourth (synchro)	3.77
Final drive	4.43 to 1
Mph per 1000 rpm in top	19.8 mph

PERFORMANCE

All figures checked to 0.5 percent by Smiths electric tachometer.

Top Speed Average	109.7 mph
Fastest Run	112.5 mph
Maximum, first	30 mph (5200 rpm limit)
Maximum, second	52 mph (5200 rpm limit)
Maximum, third	82 mph (5200 rpm limit)
Maximum, fourth	109 mph
Standing quarter mile average	18.3 secs
Fastest run	18.0 secs
0-30 mph	3.4 secs
0-40 mph	6.7 secs
0-50 mph	9.6 secs
0-60 mph	13.8 secs
0-70 mph	18.9 secs
0-80 mph	22.6 secs
0-60 mph-0	NA

	Top	Third
40-60 mph	11.6 secs	7.2 secs
50-70 mph	13.0 secs	7.7 secs
60-80 mph	12.6 secs	9.1 secs

Fuel consumption, cruising	32 mpg
Fuel consumption, overall	26 mpg

PORSCHE

FRANK COGGINS
EDITOR

Now we are going to talk about a car! You'll find our conversation larded with a lot of seemingly extraneous facts from here on but we think this will give you an idea of the rabid, total enthusiasm that owners have for this automobile.

We found, for example, that few Porsche owners let anyone else, family or not, drive their car. One girl we know shopped around in New York until she found a garage that would let her park the car herself and lock it. In Manhattan this is not easy! Porsche ownership is almost like a cult and before we say anything about the model we roadtested we freely admit to a strong susceptibility to this automobile. We could very easily be one of those "Keep your cotton-picking hands off my Porsche!" owners and will not swear that we may not own one yet. We don't have accurate figures on this, but we've found that many of our VW owner/friends are also Porsche owners and that many Volka owners buy up to Porsche. VW dealers consider dualing with Porsche (handling both cars in the same agency) a real natural.

In any event whenever we get our hands on one for a long weekend we just figure in advance that we're in for a lot of sheer joy of driving good, exotic machinery and the 356 C/1600 SC was no exception.

First of all, the minute you just sit in the car you feel that you are ready to go places. The seats are among the most comfortable obtainable in any automobile. Adjustable to fully reclining with loads of forward and backward movement, it's almost impossible not to get a good driving position. The gear shift falls right to your hand and has that solid feel through the gears that has become legend with Porsche owners.

Our test car had the 107 horsepower engine and frankly this package would seem to be the ultimate particularly with the new 911 just around the corner. Surprisingly, we got some great kicks out of the car, of all places, right in the midst of Manhattan traffic on Third Avenue during the rush hours. Here, the lovely, quick steering and that fabulous acceleration took so much of the bore out of working our way uptown that it was about the best thing a salesman could do to push the car to a prospective customer.

We are not going to go into the history of Porsche too deeply this time around—we're saving this for our story on the 911. You get bits of it in our VW stories every month in any event. The name Porsche keeps cropping up whenever talk about cars-at-speed occur.

Now, one of the things that really bothered us somewhat, and we mention this in connection wtih our MGB story, was the dashboard. The top of it is padded, which is great but why the painted metal in an expensive automobile such as this? Your gauges, of course, are well placed and can be read with no effort through the steering wheel (three spokes). The grab handle is worked

in nicely but I can't help admitting disappointment about this one facet. The upholstery is excellent and you can't beat the Porsche top for construction (we drove the Cabriolet, too).

The four speeds forward are all synchronized which helped immensely in that mad, mad New York traffic. On the turnpikes, of course, the Porsche comes into its own aided and abetted by by those powerful discs on all four wheels.

You can power your Porsche with either an 88 or 107 horsepower engine. In normal use, one of the men at Porsche of America felt that the average owner could tell not the difference and, further, that the "C" (88 hp) could outdrag the "SC" (107 hp) for a couple of car lengths. Then, we hasten to add, the SC leaves the C and takes off for parts unknown. Now this may be heresy to some of the Porsche purists but we mention it just to keep people from settling for the 88 horsepower engine and then wishing they had gone for the 107 horsepower mill.

Accessories for the Porsche include luggage racks, exhaust extensions, radios, mirrors, safety belts, floor-mats and of all things, an electric sunroof. Prices start about $4200 East Coast POE. If you feel that this report sort of teased you—that you want to know more about the Porsche we highly recommend that you get over to your nearest dealer and have a demonstration drive. You may find yourself deliberating about whether or not you want the detachable hardtop for the Cabriolet you've ordered.

The name, incidentally, is pronounced like the girl's name: Por-tia and, readers, we dig the Porsche—it sure does sort of get under your skin! •

PORSCHE OWNERS REPORT
By Trudi Dembi

When our Porsche was about six months old, we went on a business trip to Washington, D. C. and my husband drove downtown to his appointment; double parked; hopped out and blithely said: "Call for me here at four o'clock", disappearing before I could protest. In that unfamiliar city, with my heart quaking, I strapped myself into the driver's seat. By this time, I had years of VW driving experience, so that I was not helpless, but I never expected that within five minutes I would be kicking myself for not having tried the Porsche sooner. I can still remember how quick and how pleasant my adaptation was. The only thing I had to restrain was my right foot, which had become heavy on the throttle pedal from pushing VWs through the gears but otherwise I felt immediately as though the Porsche was my friend.

According to the market research wiseacres, such phenomena as three-tone automobiles and similar vulgarities result from the fact that women in the United States have an ever-increasing influence on the selection of automobiles. These "M.R." guys have never met my husband. I have never been given a chance to express an opinion as to the timing, or the selection of make or model of the family car either in the days when there was one, or now that we have more than one at a time, and one is for my use.

The first car we ever owned was a 1948 Prefect, then a 1950 Plymouth with manual gearshift, on which I learned to drive, then an MG-TD; thereafter two Morgan Plus 4-TR-2 four-seaters, with a second-hand VW acquired for and assigned to me.

Torn between his undiminished love for the Morgan and his desire, on occasion at least, to have his family ride with him, Husband was brainwashed, and after a couple of mistakes, acquired a 356B Porsche convertible, which pleased all of us more or less. Husband concedes that it is more comfortable than his lamented Morgan, but grumbles that it doesn't corner as flat. Besides, he keeps reminding me, the rough ride in the Morgan was an exaggerated complaint and disappeared at speeds of 80 and over, which is possibly true. The kids are bigger, and their legs are much longer, and they drive their own cars now, but they are very happy to borrow the Porsche on the rare occasions when husband will trust them with it.

Like a burnt child wary of fire, I had avoided driving the Porsche when it was new because of my remembered dread of the hairy Morgans. Nor was I put at ease by being growled at during the maiden voyage to keep my dirty hands off the pure white headliner inside the convertible top. I was more than content as a passenger and probably would never have taken over the controls voluntarily.

I was already familiar with the luxurious little touches which distinguish this marque from all other cars from the passenger's standpoint. I can rhapsodize about the seats, for example. The passenger seat can be moved backward and forward and the angle of the backrest can be anything from straight up to sleeping

position without interfering with the driver. The seat is also deep enough so that with the top down and window raised my hairdo survives more or less. Visibility is ideal, even for a 4′ 11½″ Brunhilde like me. The heater controls are flexible, like a dual control blanket, so that husband can direct most of the flow from him, who always wants less, to me, who never has quite enough in any other car.

I remember being stunned by the all-of-a-sudden recognition of the pleasure of driving the Porsche. It turns at a thought. Shifting up or down is effortless and one finds oneself playing with the gearbox for the fun of it. My confidence in driving in traffic, on the highway, yes, and in parking, too all overwhelmed me within the space of the first few minutes that beautiful day.

Since then I have had occasion to wonder if it was wise for me to declare flatly that I was perfectly content to drive VWs and leave the sportscars to The Man. When I figure my way out of that cul-de-sac, the next thing I will have to work on is a deeply seated prejudice of husband's. Whenever I say to him, "The next time we rally, you navigate and I'll drive," he always says: "No! If you funk out while navigating, all we lose is a few points, or at most, the rally, but if you're driving, we could lose the car." How do I convince him that even in the excitement of a rally, I couldn't get rattled driving the Porsche? Cheetahs never stumble. ●

The drophead models are fairly rare and must be checked very thoroughly for rust, as this affects chassis stiffness Hood is dead easy to operate

The driving position is very good for most sizes of driver with the pedals properly placed for heel and toe operation. Seat base sags with old age

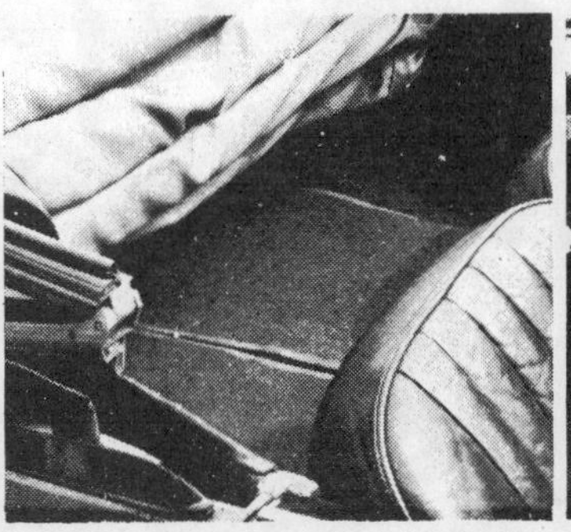

The space behind the seats is really only suitable for luggage, although you could take a passenger sitting sideways or children on short trips

The dashboard layout is simple with tachometer f as standard. There's no w temperature gauge of co but there's one for oil

Front boot space is severely limited by the fuel tank and spare wheel and can only really be used for a small bag or two. Lights aren't too good (6 volt)

The engine looks similar to a Volkswagen, but the power is slightly different! Everything is accessible for maintenance and unit comes out in minutes

Rear suspension is by swing axle system and can be improved by certain modifications. The best all-round tyres for the car are the German-made Dunlop SPs

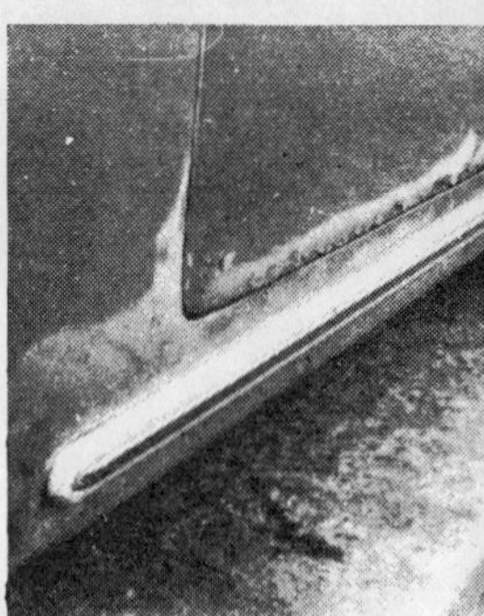

The car is a monocoque, so rust is serious on the u side. Look around the ja points. However a little rust as here is not se

GO EXOTIC!

BRIAN SMITH SUGGESTS

PICK

PORSCHE CHECK POINTS

- ***Beware a damaged exhaust system—a new one costs £25.***
- ***Body is monocoque, so rust underneath is serious.***
- ***Panels are welded in so damaged ones are expensive to replace.***
- ***Engine may not be the one stated in log book.***
- ***Convertibles must be rust free.***
- ***Engine and transmission are expensive to repair or replace.***
- ***Oil leaks indicate serious trouble.***

BELIEVE IT or not, they were Special Builders! Yes—only 22 years ago there was no such thing as a Porsche car at all. This small firm started up producing 1100 cc hand-built specials and now what do we find? In the past few years Porsches have won just about every major sportscar rally and race in the world, including the World Sportscar Championship. So how's that for progress! *('Blossomed like a Lotus' did I hear you say?—Ed.)*

These days the Porsche models are counted amongst the world's most desirable cars, not so much for their sheer top speed, but for their amazing race bred road holding and general high standards of finish. Now I can hear you saying 'but how can I afford one of those? They cost over three grand new.' Well hold on dad, cos we're not going to be talking about the very latest models, but the pre-1966 ones, which may not have the latest body styling, but are still pretty fair old cars and can be bought for under £1,000 today.

But you can find a good 356B for as little as £400. And if you've only got £200 you could get a grotty 356A—though expect trouble and expense at that age and price!

HISTORY

The name Porsche (pronounced Por-sha) was well known in the motor trade long before there was ever a car of that name, back as far as 1900 no less! Dr Ferdinand Porsche was an engineer with an international reputation and designed many cars, from the monster Auto-Union rear-engined racing cars in the '30s down to the Volkswagen dammit!

But it was his son Ferry Porsche who designed the first car carrying their name (and every one since) and who started the firm in 1949.

The first model was an 1100 cc handbuilt car with an aluminium body and set the pattern for all future Porsches. It had what was then considered a streamlined body and a horizontally opposed air-cooled engine mounted behind the rear transmission. Being handbuilt not many were made.

The first production cars appeared in 1950 again with the 1100 cc engine, which at under 50 bhp d not exactly give sparkling perform ance! In 1951 the firm offered 13C cc and 1500 cc engines as altern tives and in 1952 there were spo versions of both also available, th most powerful of which gave abo 55 bhp (DIN rating). In 1955 t 356A model was introduced with 1300 cc or 1600 cc engine in eith standard or super form, the standa 1600 cc giving 60 bhp and the sup 75 bhp. In fact the faster car becar known as the Porsche Super rather than the 356A.

In 1958 the 1300 cc engine w dropped due to lack of demand a in 1959 the 356B model was intr duced, with the choice of 60 or bhp 1600 cc engine. The car w similar to the previous model, b with various mods, including slightly different body shape.

In 1960 a 90 bhp version becar available (Super 90) which ga a top speed of 110 mph and t B model then continued with or minor mods for three years. La B models do have larger front a rear screens and twin rear grill In 1963 the last of these models w introduced—the 356C with a choi

A PORSCHE!

)f 75 or 95 bhp engines and this was the first one to have disc brakes as standard. The car was discontinued n 1965 when the 911 and 912 range was introduced.

In addition to these volume production models Porsche also made a imited number of Carrera versions. These had a double overhead camshaft, twin-plug, dry sump racing engine giving much superior performance, and were made in 1500 cc 1600 cc and 2 litre capacities. The 2 litre model is now particularly valuable as only 128 were ever nade and there aren't many of them n this country.

MODEL TO BUY

Firstly we wouldn't advise buying a pre-356 model (1955), as they were rather underpowered and by now will almost certainly be in poor general condition.

The 356A model is reckoned to have been the best body aerodynamically, but there are few of these around in good condition. either. The 356B is probably the earliest you should buy and obviously the later model if you can afford it. The price of a good 356C will vary between £700 and £1,000 according to year. There are few Carreras in this country and anyway they are competition cars really and not much suited to town use. The more powerful of the standard engines obviously costs more, like about another £100 for the 75 bhp and another £200 or so for the 90. Personally, we'd go for a 356C Super 95, but that's being greedy!

POINTS TO WATCH

Firstly let's say that buying a dud Porsche can be a heart-breaking and expensive business. Engine and transmission parts cost a lot of money and if you're unlucky you could spend as much again on repairing the car as you paid for it! One sure way to avoid this is to get Porsche themselves to check any used model you're thinking of buying. They have the history of most models in the country anyway, but will also check over the car and report verbally for a fee of £3.15 or in writing for £6.30. It could be a cheap way to avoid an expensive mistake. The address is *Porsche Cars Great Britain Ltd, Falcon Works, London Road, Isleworth, Middlesex.*

If you are going to trust your own judgment, here are the points to watch. Firstly ensure that the car

CONTINUED ON PAGE 94

Is a 10-year-old Porsche still sporty?

Porsche 356SC meets Superbug L

NSW
PM-044

IT'S FUNNY — whenever we tell anyone we've just done a comparison test between a 1973 VW Superbug and a 10-year-old Porsche 356 SC they assume that we've done it to discredit one or other of them.

They realise the two cars' prices are now vaguely similar and assume we're going to compare them directly and unfeelingly entirely on a cost basis.

VW enthusiasts bristle because they think we're going to imply that their cars are 10 years out of date.

Porsche pundits would simply prefer to forget that the other two-door, 1600 cc, air-cooled, rear-engined, all independently suspended car exists.

So let's get it right from the start. Our idea for a comparison grew from our respect for both cars. The Superbug has an old, old body, but its suspension and brakes are first-class and its reliability is legendary.

The Porsche was a brisk but not sensationally fast car in its day. Its main plus features were a fine cruising ability, reliability and the satisfying, accurate way it could be driven.

Our test Porsche belonged to a young systems analyst, who had carefully righted the wrongs of several previous owners and was maintaining it in mint condition. The Porsche was so good we found it difficult to believe that it was used in the peak hour traffic every day and raced in club events on weekends.

Its solitary non-standard feature was an Abarth exhaust system which projected four elegant pipes under the rear bumper, but reduced the ground clearance quite a lot.

The VW was one of Volkswagen Australia's test fleet — a Superbug L with radials. That's the sportingest beetle made.

The mechanical similarity of the cars is obvious. Their engines are the same size and related, though the Porsche is in a much higher state of tune.

Both had the same whirring, muffled mechanical noise coming through the rear panel of the cabin and the whine of the cooling fan as the revs rose.

Here was one of the main dividers between the cars. The VW's revs rose — but not far. It tended to run out of breath well before valve bounce and revving out beyond a certain point became noisy, laborious — and slow.

By contrast, the Porsche wasn't all that strong down low, but it really sang up to its redline of 6000 rpm with power all the way.

The gearboxes had a similar feel, though the Porsche change was quite sloppy by today's standards. The syncro was just as quick as ever though.

The VW had, if anything, a worse change than previous models have had. There seemed quite a wide gap between the gear planes and its action seemed a bit "loose", too. Syncro was fine.

The Porsche had far and away the more aristocratic exhaust note — a deep, smooth throb. The Beetle's exhaust was almost entirely overshadowed by mechanical and fan noise.

Both cars had a robust, unmistakeable "Germanness", though in 10 years between them vehicle builders have graduated from mainly metal appoint-

ments to first quality plastic and leatherette (as used in the Superbug), the same quality was evident in both.

Both cars had floor pivoting pedals and firm, wide bucket seats with rather upright backrests.

There were similarities in the way the cars handled, too. Both had a very little initial understeer, but this could be converted to mild oversteer by applying the power on exit.

The Volkswagen's graduation to oversteer was much more predictable than the Porsche's, it gave quite a lot more warning — showing off 10 years of suspension sophistication. The Porsche, of course, had a swing axle setup and this particular one had a camber compensator.

Make no mistake about it, the Porsche was a quicker cornering car than the Beetle. Its cornering limits were impressively high, but we weren't keen to approach them in confined spaces.

The VW was much more of a "hang the tail and get it back" machine. But it had more body roll and a softer suspension than the Porsche and tended to lurch a bit through ess-bends.

Steering systems felt "related". Both cars had fairly large, slightly offset wheels set close to the dashboards and they were both quite high geared and light as you'd expect from rear engined cars.

The Porsche steering didn't seem to be 10 years old; it was very accurate and transmitted a great deal of road feel. The VW was perhaps a bit more rubbery, particularly close to the straight-ahead position, but still very good.

Things were different under brakes. The VW with its disc/drum set-up had much more initial bite than the Porsche. Its whole braking behavior was strong, stable and fade-free.

The Porsche brakes felt rather dead by today's standards, though they were perfectly safe and progressive. The position wasn't helped by the poor pedal layout though.

The Porsche surprised us with its good ride. It wasn't all that level, but it was quiet and the bumps were absorbed in a "imperturbable" way. Wheel control was fine — no skittering about over bumps even though the suspension settings were definitely "sporting".

And there was an accuracy about the car's behavior which we loved.

We found out that the Porsche still wasn't disgraced in today's company — it had maintained its viability as sporting transport and still had an adequate turn of speed.

The VW proved far better than we'd thought it might. It was an enjoyable car to drive — even though it has no sporting pretentions — and we liked its overriding strength and robustness.

We haven't any illusions about the body design — it's *at least* 10 years out of date — but we'll still regret it when the bug finally does leave the market because there'll probably never be such a robust, reliable car again for so little money.

For the beetle we feel affection, for the Porsche, admiration. Neither feeling affects the other.

—STEVE CROPLEY

PHOTOGRAPHY: UWE KUESSNER

Here's a nice clean one—used price was £450

PICK A PORSCHE!

DATA

SPARES (Super 75)

Exchange pistons & barrels £52.
Exchange pistons & barrels (90) £87.
Exchange clutch complete £17.
Exchange brake shoes (4) £8.
Brake pads front pair £4.60.
Standard silencer £25.
No exchange engine or transmission.

PERFORMANCE (Super 75)

0-60 *mph acceleration* 12 *secs.*
Maximum top speed 105 *mph.*
Fuel consumption 30 *mpg.*
Last new price (1965) £2,064.

CONTINUED FROM PAGE 89

actually has the engine stated in the log book. Don't trust the name badges fitted to the car! The standard 1600 cc engine has a black fan housing, but the Super 75 and 90 versions have a grey one. The standard 60 bhp and Super 75 bhp have Zenith carbs, but the Super 90 has a twin-choke Solex.

The air-cooled engine makes a fair whirring noise when revved, but should otherwise be smooth and quiet. If it rattles or leaks oil, forget it. Just to buy exchange pistons and barrels costs £50 (£87 for the Super 90!). The transmission is the other expensive bit and this is normally smooth, fast and quiet. It's an all-synchro unit and lack of synchro may indicate an inconsiderate previous owner. If so, make sure it doesn't whine excessively or jump out of gear.

As with any used car you should look for rust, particularly if it's one of the few drophead versions which are around. The body is a monocoque and all panels are welded in so repairs are expensive. If there's rust in the door sills and around the jacking

. . . so panel repairs are expensive!

points it's probably not worth buying.

All Porsches are undersealed from new so suspect accident damage or rust repairs if there are any clean parts underneath. If the body has been bent in an accident it has to go back to the concessionaires and go in the jig, so watch someone else drive the car down the road and if it's travelling sideways in a straight line, say thank you nicely and leave!

MAINTENANCE

The most important thing about servicing a Porsche is that it's done exactly according to the handbook and at the recommended intervals. The factory don't like owners to do-it-themselves, but in fact there's nothing very complicated about the car. Three points are particularly important. Firstly no additives must be used in the transmission oil. Secondly don't forget to change the oil filter (which is under the engine) at oil changes. Thirdly you mustn't forget that the twelve (yes 12) grease points on the front suspension must be done every 1,500 miles or more frequently in really bad weather! If you neglect the steering link pins they seize up.

Whilst all parts are fairly expensive the overall running costs need not be exorbitant. The car can do 35 mpg if driven moderately and the engine only uses ordinary diesel oil! Brake squeal on drum braked models is unusually nothing more than excessive dust in the drums.

All models have a 6 volt electric system and the headlights tend to deteriorate in time. The Carello quartz iodine conversion is worth fitting for fast night driving.

TUNING

On the road the Porsche performs pretty well even as standard, but before discussing any tuning it's worth noting that the tachometer, which is a standard fitting of course, has a special marking on it. Like any other car there is a red area indicating the maximum permissible revs. In fact with this engine you can go into the red area briefly in the gears without doing any damage, but sustained cruising in the red will wear out the engine very rapidly. However below the red area there is a large section marked in green. This indicates the useable rev range. It's a cammy engine and if you go below the green it's not very tractable, so the advice is to keep the needle in the green bit!

Bearing in mind that the engine is tuned to a fair extent as standard, there is a limited amount which can be done to improve it. Engine performance, fuel consumption and reliability can be considerably worsened by incorrect ignition timing. If you so much as adjust the contact points you *must* have the timing reset *static* to 6 mm btdc on the standard engine, 6½ mm on the Super 75 and only 3½ mm on the Super 90.

You can convert a standard engine to 75 bhp spec, by fitting exchange pistons and barrels, and the appropriate cam, followers, valve springs and carburettor jets. This would cost over £100 in parts and you'd be well advised to let an expert fit them.

It's not practical to convert the 75 bhp engine to 90 bhp spec as there are too many differences. If you want the most powerful engine, you'll have to buy one new (very expensive) or secondhand (very rarely available). It's better to buy a Super 90 car in the first place!

All the engines do benefit from gas flowed ports though, and one of the few specialists in this is *Chris Maltin, Braywood Farm, Hawthorne Hill, Nr. Maidenhead, Berks.* Otherwise you are best advised to leave the engine as standard.

SUSPENSION

Being a rear-engined car it is prone to oversteer, and there is a certain amount you can do to counteract this and also improve the general road holding a little. The front suspension can be lowered a maximum of two inches on the adjustable torsion bar. The rear can also be lowered by fitting shorter springs. Adjustable dampers, such as Konis, can be fitted and these are best set to the hardest setting on the front and softest on the rear to give better rear end grip.

Late B and all C 356 models already have softer rear torsion bars as standard, plus a compensator bar, so the damper setting probably would have less differential. As with many cars there are tyres to suit the Porsche and in fact German made Dunlop SPs perform best and these are available from the concessionaires. Standard wheels are wide enough for road use on road tyres.

The model is not a competitive race or rally car, but people do race and rally them for fun.

The Porsche is a quality car and you therefore pay quality prices both for the car, spares and service. And of course insurance can be expensive, although some companies class the fixed-head models as saloons. But it's reliable, economical and quite out of the ordinary. It grows on you!

USED-CAR PRICES

approximate retail prices for cars in average condition

MODEL	1962	1963	1964	1965	1966	1967	1968	1969
AR Giul T1 Sal	—	305	340	410	500	670	775	985
AR Giul coupe	—	—	595	735	895	1045	1600	1745
Austin A40	170	195	235	275	315	365	—	—
Austin A60	230	275	325	385	450	530	620	725
Aus Heal Sprite	245	265	300	375	440	500	590	695
Aus Heal 3000	390	460	565	670	810	965	1095	—
BMC Mini	150	190	225	265	310	370	430	500
BMC Cooper	180	220	265	310	365	425	485	570
BMC Cop S1071	—	225	265	—	—	—	—	—
BMC Cop 1275	—	—	320	370	440	515	600	700
Citroen DS	215	275	350	425	700	865	1090	1335
Daimler Dart	350	425	510	—	—	—	—	—
Fiat 500	130	155	180	210	245	285	330	385
Fiat 600	135	160	195	230	265	310	360	420
Ford Anglia	140	170	200	230	265	305	355	—
Ford Escort 1300	—	—	—	—	—	—	580	650
Ford Escort GT	—	—	—	—	—	—	615	700
Ford Cor 1200	185	215	250	290	420	480	565	660
Ford Zeph 4, V4	205	245	285	335	450	515	625	735
Ford Cortina GT	—	270	320	370	430	570	660	770
Ford Cort Lotus	—	325	390	460	540	670	785	920
Ford Zodiac	215	265	315	365	415	670	810	970
Hillman Imp	—	185	220	285	325	380	440	510
Hillman Hus Est	—	120	145	175	210	350	435	525
Hillman Minx	150	200	270	350	395	495	580	675
Jaguar 2.4 Mk2	285	350	425	515	635	840	1035	1245
Jaguar 3.4 Mk2	295	365	455	550	680	960	1085	—
Jaguar E-Type	590	710	855	1020	1210	1450	1740	1895
Merc 190 & 200	360	440	530	635	750	900	—	—
MG Magnette	230	280	330	395	465	550	645	—
MGB	355	405	465	545	640	745	860	975
MG 1100	220	255	305	360	420	505	625	—
MG Midget	250	270	300	380	440	505	595	695
Morgan 4/4	290	335	400	475	565	695	780	1025

MODEL	1962	1963	1964	1965	1966	1967	1968	1969
Morgan Plus 4	360	415	475	560	655	760	—	—
Morris 1100	190	225	265	310	365	455	505	590
Morris Minor	210	235	270	310	360	415	480	550
Morris Oxford	230	275	325	385	450	530	620	725
NSU Prinz 1000	130	160	195	235	285	345	410	480
Peugeot 403	170	200	245	325	400	—	—	—
Peugeot 404	210	275	350	440	540	655	780	940
Porsche 1600	520	610	805	935	—	—	—	—
Reliant Sabre 6	—	120	160	210	—	—	—	—
Renault R4	155	180	215	260	310	365	410	475
Renault R8	—	150	195	245	295	390	480	570
Riley Elf	175	200	260	310	370	420	495	580
Riley 1.5	170	205	245	290	—	—	—	—
Rover 2000	—	455	510	620	750	915	1095	1280
Rover 3 litre	255	335	410	550	790	940	—	—
Saab 96 & V4	145	175	210	245	445	510	600	695
Simca 1000	140	180	220	270	305	355	515	565
Singer Gazelle	175	220	285	380	420	545	640	740
Singer Vogue	275	305	370	475	545	615	715	830
Skoda Octavia	—	115	145	180	220	265	320	410
Sunbeam Alpine	255	300	365	450	530	625	740	—
Sunbeam Rapier	225	280	340	400	435	775	875	1020
Triumph 1200	200	230	270	310	360	415	485	560
Tri Herald 12/50	—	235	275	320	370	425	—	—
Triumph Spitfire	240	280	315	380	455	530	650	740
Triumph 2000	—	—	465	560	665	785	925	1080
Tri TR4/TR4A	365	415	465	515	570	660	770	—
Triumph Vitesse	225	255	300	355	420	510	650	785
Vauxhall Victor	185	215	250	310	360	420	620	720
Vauxhall Velox	165	210	260	310	—	—	—	—
Vauxhall Viva	—	185	220	255	300	435	505	585
Vauxhall VX490	230	275	330	400	460	535	—	—
Volks 1200/1300	245	275	320	370	420	480	555	635
Volvo 122	310	380	460	555	665	765	—	—
Wartburg	—	—	135	160	190	230	—	—

The shape of things to come

Porsche's post-war 356 model set a style from which the company is only now starting to depart

MENTION the name Porsche to a motoring enthusiast and the image conjured up will be of a road burning 911 or maybe a smooth sophisticated 928, or even one of the Le Mans 917 racers.

Mention the name to someone interested in the history of the car and the image will be of a remarkable father and son, two of the most brilliant automotive engineers the world has ever seen. The story began in 1875 when Ferdinand Porsche was born in Bohemia. Even at an early age he was interested in things mechanical and quickly showed a remarkable creativity in a pioneering age. By 1906 he was made chief of engineering and production at Daimler.

In the years that followed Professor Dr Porsche was to stamp his name firmly on automotive history. Among the many cars, engines and military vehicles he was to design were the all-conquering Auto-Union racing cars of the '30s, the Mercedes SS of the '20s and, perhaps most famous of all, the "people's car" — the Volkswagen.

His son, Ferry Porsche, was born in Austria in 1909. He too, was quickly to show an interest in engineering and in 1930 became a designer in his father's consultancy in Stuttgart, and five years later became director of the test department in the new Porsche works at Stuttgart - Zuffenhausen. Ferry Porsche was fortunate that his father had spent many years working for other companies, for throughout his career Ferdinand Porsche found his radical ideas coming up against brick walls, or at least up against financial brick walls. Eventually he decided to work for the one person he knew would appreciate his ideas ... himself.

Thus it was in 1930 that father and son joined up to form the engineering consultancy that was to go down in history — and produce the most popular car in the world, the Beetle. And it was the Beetle that directly spawned the Porsche 356, that in turn led to the 911, 924 and 928 of today.

They came up with the blueprint for a simple, light car to provide reliability, performance and economy. The line of thought is easy to follow — as the price was to be very low, that would mean most prospective owners would not have a garage, so the engine was to be air cooled to avoid freezing in winter. Costs would be cut if the engine and transmission were near the driven wheels, so the finished product was to have an air-cooled flat four engine of around one litre at the back of the car.

But while it was easy enough to put down on paper, backers were more difficult to find. But then along came Hitler and he backed the idea of a people's car ... Volkswagen was to be powered by a one litre engine developing 26 bhp and providing a cruising speed of just over 60 mph. Oh yes, and it was to cost next to nothing!

The year was 1934 and the project had finally been given the go-ahead. The first three were built by hand by Porsche and his workers and before long, some sixty development cars had been produced built by other manufacturers like Mercedes, under Dr Porsche's watchful eye, of course. But no sooner had the teething troubles been ironed out, the development team satisfied and the dream become a reality, than the war came and the project was shelved once more. The only Volkswagens to see the light of day were the development cars and the VW-based Kubelwagen Jeep-type vehicle.

Now as the world returned to near normality and production of the Beetle finally got underway Ferry Porsche was able to look to producing a car with the Porsche name. His father, weakened by imprisonment after the war, was too ill to take part in the plans himself but gave the project his blessing. It stood to reason that the proposed sports car should be based on the Volkswagen: after all that was just about the only source of bits around.

Indeed the company was so hard up that most of the bits they could not salvage from a Volkswagen had to be begged, borrowed or smuggled into Austria where Porsche were now based. The very first car to bear the Porsche name was a real 'bitsa' car, made up from bits of this and bits of that. It was a roadster because that needed less metal to make and the slab-sided body was formed of alloy panels around a strong box-section frame. Mechanical parts came from the Beetle although judicious work on the cylinder head raised the compression ratio to 7 to 1 and improved carburation boosted the power to 40 bhp. Swing axle suspension was taken from the Beetle and the resultant car weighed just over 11 cwt and had a top speed of around 90 mph ... a good deal quicker than the Volkswagen.

Because of the lack of funds the car was promptly sold and the money used to make another and so it went on. It was 1948 and these cars proved to be the start of the 356 line, and a line of cars that was to last a full 17 years. Even as the last of the 78,000 356s rolled off the production line in 1965, it still bore a strong resemblance to those early cars of the late 40s. Change for change's sake has never been a Porsche trademark, although like the Beetle, the first and last of the 356 line were substantially but subtly different.

Power was increased eventually to 95 bhp, speeds topped 110 and the engine size increased to 1582 cc. The basic shape of the little projectile remained much the same, though even that had changed — subtly — over the years. Windows got

WHAT CAR? RARITY/PORSCHE 356

In the beginning was the Volkswagen and it was from such humble origins that the smooth and slippery shape of the Porsche 356 evolved

Steering wheel would be unfashionably large in a modern car

With a rear engine the front 'boot' holds the fuel tank and spare

Following the VW trend, the 356 has its power pack at the back

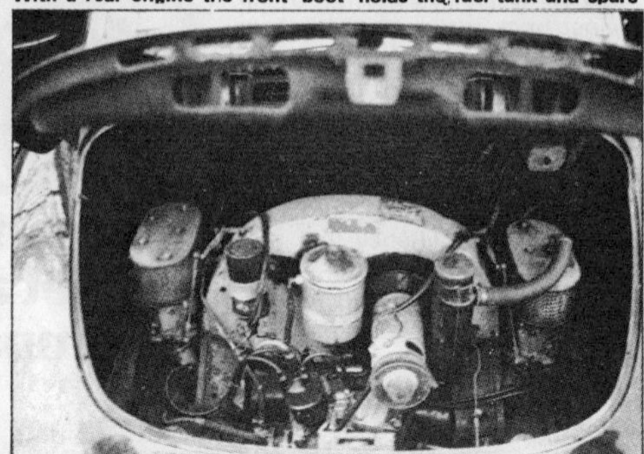
Air cooled flat four engine gives the later models some 95 bhp

larger and the lines became sleeker.

When the car pictured here first took to the English roads in 1964 the end of the line was in sight for the 356. The previous year had seen the public debut of the 911 at the Frankfurt show, and in 1964 production of that car began — it's still in production and in great demand today.

Our rarity is a 356C with the uprated SC engine. In many ways its specification is quite advanced — all-round disc brakes do an excellent job to stop the little car — and its performance will shock much newer 1.6 litre cars. The SC has a 9 to 1 compression ratio, light alloy cylinder barrels, twin Solex carburettors and 95 bhp. Neat touches include a lockable fuel filler — released from within — and variable speed screen wipers; the car even has a cigarette lighter. A pair of tiny rear seats turn it into a 2+2, just.

Dating features are aplenty, too. The huge steering wheel is a bit of a handful though the steering is delightfully light and direct as it ought to be with all the weight at the back, and the six volt electrical system lacks some of the punch of more modern 12 volt systems.

The car's owner is Kings Road fashion designer Anthony Price who admits he first bought the Porsche on the strength of its looks. He said: "It's such a beautiful shape that I enjoy just looking at it. But of course being a Porsche it's a delight to drive, too."

On the road the roar of the flat four just behind the driver's ear and its slightly uneven beat help to create an impression of power that matches its performance on the road. Contemporary road tests gave it a 0-60 mph time of 13.2 sec. But this is no temperamental sports car, for as its heritage would suggest, the 356 is a car that is prepared to be driven under all sorts of conditions. With its superb steering, excellent gearbox and willing engine the 356 can simply be entered, switched on and driven away. Anthony Price uses his car to commute to his Kings Road shop every day and the car shows no signs of being upset by the routine.

He said: "Although it does not mind running around London all week, the car really does look forward to a long run: over 90 mph, there's just a little rattling to worry about."

Perhaps the only question mark over the car when new concerned the handling. The ride is excellent for a car of its vintage and pretentions but those Beetle based swing axles had a tendency to rather sudden breakaway if treated badly. These days it's a problem that the driver is aware of and caution is the watchword.

Owning a car such as this, especially one in constant use, does have its problems. Anthony said: "I've had the car for three years and have given it one respray but the time is coming up for another one as there are a few rust bubbles showing through in places. But my biggest worry is spare parts for they virtually don't exist. The last major part I had to replace was a fuel tank — I got the last one the garage had."

But it must be said that being a Volkswagen at heart, it is not often the little black Porsche does go wrong ... that engine is almost unburstable. The car is reliable solid and dependable just like every Beetle that trundled out of Wolfsburg. But unlike the people's car, the 356 has speed and style. And as its owner says ... "It's just the sexiest looking car ever made."

The original steel streamlined body appears to be somewhat dumpy-looking these days.

BUYING USED

The Porsche 356

TONY LEWIN discovered that you can become a Porsche owner for only three figures.

"What's in a name?" a famous literary figure once made one of his tragic characters ask. Sure as anything, the bard's immortal words are just as relevant today, and can even be persuaded to apply to buying certain sports cars on the second hand market.

For Porsche enjoy a position among a select band of makes such as Mercedes Benz, Rolls-Royce and Aston Martin which apart from their double barrelled titles also have in common another quality; they are what an advertising man might describe as 'aspirational' cars — in everyday language, makes of cars that people feel they want to own.

The ambition to own a Rolls-Royce — irrespective of model or year — is a common, if less often realised one. So it is with Porsche, too. Just to be a Porsche owner is to be elevated several rungs on the social ladder, and to be able to point nonchalantly at the best bitter pump in the saloon bar with a bunch of Porsche keys is more impressive still.

But who's to know you only paid a three-figure sum for your example of one of the world's most desired motor cars?

They're around for that sort of money in the form of the original and long-running 356 series, predecessor to the current 911 model which is only now beginning to be threatened with extinction.

Any Porsche for under a thousand is clearly a status bargain, but is the 356 really such a financially painless route into the Porsche club as it seems?

True to its Volkswagen ancestry, the first Porsche changed very little externally in its seventeen year production run but, again like Volkswagen, a 1948 car and the final 1965 model placed alongside one another would amply illustrate the development that had taken place over the years.

The air-cooled flat four Volkswagen engine of 1948 (which even then produced the directly un-VW like output of 40bhp) grew steadily from 1100cc to the full 1600 of the final versions tuned to give around 95 ultra-reliable horsepower.

If the rules are stretched somewhat and we consider the VW engine simply as an air-cooled flat four, then the ultimate development of the 356 can be regarded as the two-litre Carrera with its four bevel-driven overhead camshafts and full 130 hp output.

Variety of engines

Once again, the external similarity of the cars is deceptive. Under the then-streamlined but now rather dumpy-looking steel bodywork (only the earliest cars featured aluminium) lurks a bewildering variety of engine types and sizes.

Early cars had one single-choke carburettor per bank, later ones twin-barrelled instruments. As the engine gradually increased in capacity, various states of tune continued to be offered. In general, the 'S' or Super versions developed their greater power thanks to different camshafts and larger valves, though Porsche were always playing tunes on cylinder barrels, pistons and crankshafts, too.

The Super 90 is perhaps the best known in this country, but a Super 75 and a 'standard' were also offered at the same time, and the version, coinciding with the introduction of the mildly revised 'C' series body, was the 1600SC with yet more power — 95 bhp.

At the peak of its glory in 1963-64 the 356C was an exclusive car indeed, but as with present examples of the marque, it was far from cheap and had in consequence a rather elite following. It cost some £1,000 more than the E-type Jaguar of the day, and was only able to offer a maximum speed of around 110mph, as opposed to the big Jaguar's easy 150 plus.

By modern standards its acceleration performance was unimpressive, too, with 60mph taking a full 13.2 seconds according to contemporary road tests, but it is often forgotten how much standards have improved. Fuel economy was the 356's strong suit — 35mpg being the rule on long runs.

Current Porches carry a six-year non-rusting warranty, but if the 356 car had had the same cover it is unlikely that the firm would be with us today, so great would the warranty claims have been.

In the words of Acton-based Chris Turner, who services Porsches for a living, the 356 is a "rust bucket pure and simple".

The sad truth is that most 356s proudly displaying the Porsche badge are very tatty indeed, and this tattiness is hardly ever superficial. Surface rusting is in general cosmetic, but replacement body panels are very hard indeed to obtain — Chris claims he has the largest stock of original new body panels in the UK — three or four wings, a couple of bonnets and the occasional underbumper section! Rust appears in all the classic places — and more. Look around the headlamps, horn grilles, tail lights, wing edges and particularly the door latch panels. In short, look everywhere.

More important structural rust affects the battery box floor (below the spare wheel in the front), the front suspension A-bracket ends, the inner sills where they meet the floor panels and the footwells behind the pedals. Other trouble spots are the torsion bar carrier box section at the rear, the front torsion bar tube pick up points and the anti-roll bar chassis mounting points.

It's a gory list of corrosion problems, and it means that there are relatively few sound examples still in existence.

Rare spares

Replacement floor panels are available from the US and cost only £100-£120, but putting them in might run up a bill ten times as great.

Perhaps the greatest problem is spare parts availability — or non-availability in the case of the 356. Components such as brake slave cylinders, brake drums and cylinder barrels are like gold dust, as are certain body panels.

The factory no longer produce or distribute spare parts for the model, but there is a thriving and enthusiastic owners' club as well as a network of specialised suppliers — provided the owner is willing to pay freight from Germany or the United States.

In conclusion, the 356 may be an interesting car in its own right and it certainly is a convenient back door to Porsche ownership. But its rusting propensities will rule it out for many, and the scarcity of spares is a real worry. It is most definitely not the car for the non-enthusiast, but then cars with true class and character rarely are. ■

Only a couple of these very rare Speedsters remain in this country.

The rare and fast Carrera engine (4 ohc, 130 mph).

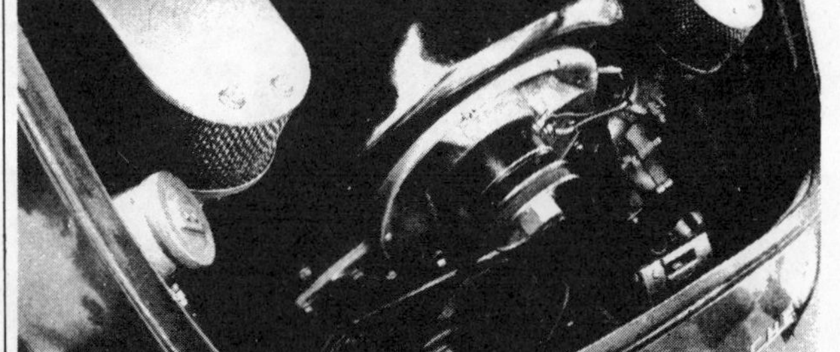

CALIFORNIA
XBJ 435

The Replica Classic

the plastic Porsche copy that's now as collected as the original

By John McDermott

WHEN he started designing cars in the small village of Gmund, Austria in the immediate post-war period, Ferdinand Porsche could have possessed little idea that he was about to create a legend. Quite understandably, for it was (and still is) unthinkable that one of the world's most successful sports car marques could possibly be constructed from such lowly and unlikely ingredients as Volkswagen engines and running gear. It is a "silk purse from a sow's ear" story of such implausible magnitude that even Hollywood would be hard pressed to exaggerate and it is every bit the equal of the pantomime transmutation of a pumpkin-plus-mice into a beautiful coach-and-eight.

In their second decade as purveyors of mutton dressed as wham (for no matter what the basis of their vehicles, they were ever impressive) the Porsche company introduced a two-seater convertible which underlined the direction in which they were headed. The Speedster was a milestone in the marque's history, an instant classic and one of the most desirable light motor cars of the mid-'50s.

It was rather cubby looking, an aerodynamically lumpy yet supremely lovable vehicle which, in open-topped trim had a mere three-year life span. In 1958 the Speedster's 1582cc engine propelled the curvacious bathtub body through still air at just over 100mph. That alone was hardly headline news — it was the way it got there which fired the imagination and enhanced the then young legend, for it was every inch a driver's car, one that gave back what it was given. Piloting a Speedster was akin to flying at ground level, a feeling amplified by the wind in one's hair and the flat, beavering drone of the engine over the shoulder. Heel-and-toeing it through Alpine swervery, arms extended and cravat flapping forward in the warm vacuum of rapid progress was the stuff of dreams that filled many a young head.

One would be inclined to think that to imitate Porsche in any way whatever would be foolhardy to the point of lunacy, a surefire trip to ridicule and failure. Not so, for it has been done — and done so successfully that the copy is fast becoming as collectable as the real thing. One of the first plastic classics. Furthermore, the cessation of production at the end of this year will see to it that the ersatz Speedster's collectibility quotient rockets as never before, no doubt pushing it into the investment category.

Automobili Intermeccanica was founded in Turin in 1958 by Frank Reisner, a chemical engineering graduate of the University of Michigan. By 1974 they had produced just over one thousand cars which included Imp, Apollo, Italia and Indra models mostly destined for American customers. In addition they had built many prototypes on a sub-contract basis for other factories, the most notable being for their close neighbour Lamborghini. It was AI who built the very first Lamborghini ever put on the road and they followed this by constructing other pre-production models bearing the bull badge. Some four years ago the company was bought by Tony Baumgartner, president of a large West Coast VW dealership and moved to California. The Speedster replica business was a natural for such a set-up and work began on the new model as 1976 became 1977.

Intermeccanica's Porsche look-alike is made to order although the less busy winter period is used to stockpile cars for the Spring rush. Come early April and perhaps twenty vehicles or so are ready for immediate delivery, the rest following on at the rate of one a day.

It takes six skilled glass fibre workers to turn out one monocoque bodyshell per day, including the mated bolt-ons such as doors, bonnet, boot lid and seat pans. Constructed, in the main, with a thick, basket weave roving reinforced by heavy cloth, the bodies are colour impregnated and, where prone to vibration stress — such as at the wheel arches — the glass fibre is lipped and rolled.

The inverted shells are married to a hefty box-section steel subframe with glue and rivets before being undersealed from stem to stern. This integral strength eliminates the pan flex and subsequent metal fatigue of the original, the main blessing of which are doors that suffer no droop even after many years of use.

Three views that show the very high standards of replication.

The Replica Classic

Once dry, the bodyshell is upturned and pushed around the small factory on a barrow jig giving the workshop the appearance of a fairground dodgem ride. What follows is the gentle, almost totally silent magic of bespoke car construction, a painstaking process which seems to rely more on the trained eye and sensitive touch of the craftsman than on the tools he wields. Even the nose and short hairs on the back of the neck seem to play a part in each car's creation, every job being checked with the fingertips and a one-eyed sideways glance before being passed up the line. Yet despite this untechnological approach, real precision results. So authentically reproduced are these Speedster copies, most parts will fit original Porsches under renovation. Intermeccanica's hand-fabricated Speedster windscreen frames, for example, are today's only source of supply to collectors world-wide.

Such detail is expensive and results in a 1958 replica costing the same money as a brand-new Porsche 914 on America's home market. Tony Baumgartner shrugs in disbelief at the success of his well-forged copies. "Who'd pay that kind of money for a car with no windows? I'll tell you. He's 40 to 45 years of age and was in college when the original hit the street. He was maybe rooming with a wealthy guy who, having a sports car, always got the girl. Now he's in a position to indulge in a few 'keep young' fantasies. He's probably buying Johnnie Ray records too. Crazy isn't it? It's like a bathtub and guaranteed to leak, but I guess we're all a little crazy in our own way."

Good quality dressings are used in the Speedster, no doubt accounting in no small way for the $16,500 price ticket. With the exception of cars built for Texas, the cars are upholstered and trimmed in the best Connolly hides, even down to the leather edging around the floor carpets. Nardi supply the wood-rimmed steering wheel; Blaukpunkt the stereo; and Haartz cloth is used for the convertible top and tonneau. The Porsche name does not appear anywhere on, or in, the car. It's an Intermeccanica, not a Porsche but in all possible respects of looks and performance there is barely a ha'porth of difference. In its constructor's words "We meet or exceed all original specifications and in many ways our car is better than the Porsche. One would hope so, we have the advantage of a twenty-year period that has produced some terrific materials".

It is, of course, quite easy to match the standards and targets of yesteryear and not too difficult to exceed them, even those set by a genius such as Porsche. Comparison is, therefore, to a great extent unnecessary beyond the barest of academic interest. They are worth making for one reason: to show how evident the copier's skill is and, imitation being the greatest form of flattery, Porsche buffs should at least acknowledge the gracious compliment being paid their mentor.

Figures for the 1958 Speedster are in parenthesis, the nouveau's modified Volkswagen 1640cc twin-carb power unit being responsible for their almost perfect recreation:

Top speed/mph:	(105) 107
3rd gear max/mph:	(82) 72
2nd gear:	(53) 47
1st gear:	(30) 25
Standing ¼mile/secs:	(17·1) 16·9
0-30mph/secs:	(3·0) 2·9
0-60mph/secs:	(10·5) 9·5
BHP at 5000rpm:	(88) 90

The net result is a car which retains much of the tight, firm ride and handling of the original, the same measure of driveability and an uncanny reconstruction of the pleasures of long ago. If only one could replace the pear-drop smell of glass fibre with that of under-carpet damp it might well be a trip in a time machine and the picture would be complete.

Perhaps the most surprising feature of the deal is that prices now being asked for the copy are almost the match of those demanded by the real thing. It would be more understandable if the rising prices of the new offering simultaneously resulted in an increase to the originals, but that does not seem to be the case. Could it happen that the day will dawn when an Inter will be worth more than a Porsche?

Today's cost new is $7000 above the going rate for a new car built two years ago and whilst that itself has kept used prices up, they have not yet proved to be a real investment. But it is guaranteed that, with only a few more to be built, an Intermeccanica Speedster will soon be as good as money in the bank.

Delivery is two months ex-factory at this time, but add shipping, duties and taxes to the cost and that investment will take many years to come good for European buyers.

As for Automobili Intermeccanica, they will probably replace their present replica business by "Crayfordizing" Ford Mustangs. A dull business but at least one that will excite many and offend nobody. ●

Above, authentic side vision too! Below, spartan interior but leather seats add luxury.

Above, VW's slightly bigger twin carb engine provides more power than original. Below, a bonnet-full of spare tyre.

PORSCHE BY MODEL

PORSCHE 924 1975-1981

The 924 story is told through seven road tests, 3 comparison tests, a track test, a 12,000 m. report, a technical analysis, previews, driving impressions, new model introductions and specifications. Besides the 924, the turbo, Carrera and the Butera supercharged models are covered in detail.

100 Large Pages

PORSCHE 928 COLLECTION No. 1 (1977-1981)

This book is comprised of 18 articles, six of which are Road Tests and four Driving Impressions. They cover the model from its introduction in 1977 through to 1981 and report on the 928, the 928S, the B+B Targa and the Automatic version.

70 Large Pages

PORSCHE 911 COLLECTION No. 1 (1965-1975)

The first 10 years of the 911 are reported through 5 road tests, a comparison and used car test and a 50,000 mile report. Models covered are 911-S-T-Lux-E-Targa-Sportomatic-Carrera 2.0-2.2-2.4 and 2.7.

70 Large Pages

PORSCHE 911 COLLECTION No. 2 (1974-1981)

The sixteen articles in this book contain 5 road tests, a 12,000 m. report, a comparison test, new model introductions plus articles on preservation and racing. Models covered include the 911S —SC—LUX—4WD—Targa and Turbo Carrera.

70 Large Pages

PORSCHE TURBO COLLECTION No. 1 (1975-1980)

Five Road Tests plus 12 other articles make up this first book devoted to Porsche Turbo models. Reports are drawn from the U.S., Britain and Australia and cover both the 3.0 and 3.3 litre cars the 924, 930, 935 and 936 plus the 911 Sportomatic Turbo and the Carrera.

70 Large Pages.

PORSCHE 914 1969-1975

Road tests, comparison tests, owner surveys, service testing and specifications. Over 140 illustrations covering the 914, 914/6, 914SC, 916 and 914/2.0L from the leading U.S., Australian and British motoring journals.

100 Large Pages

These soft-bound volumes in the 'Brooklands Books' series consist of reprints of original road test reports and other articles that appeared in leading motoring journals during the periods concerned. Fully illustrated with photographs and cut-away drawings, the articles contain road impressions, performance figures, specifications, etc. None of the articles appears in more than one book. Sources include Autocar, Autosport, Car, Car & Driver, Cars & Car Conversions, Motor, Motor Racing, Modern Motor, Road Test, Road & Track and Wheels. Fascinating to read, the books are also invaluable as sources of historical reference and as practical aids to enthusiasts who wish to restore their cars to original condition.

From specialist booksellers or, in case of difficulty, direct from the distributors: BROOKLANDS BOOK DISTRIBUTION, 'HOLMERISE', SEVEN HILLS ROAD, COBHAM, SURREY KT11 1ES, ENGLAND. Telephone: Cobham (09326) 5051 MOTORBOOKS INTERNATIONAL, OSCEOLA, WISCONSIN 54020, USA. Telephone: 715 294 3345 & 800 826 6600

ROAD & TRACK ON PORSCHE

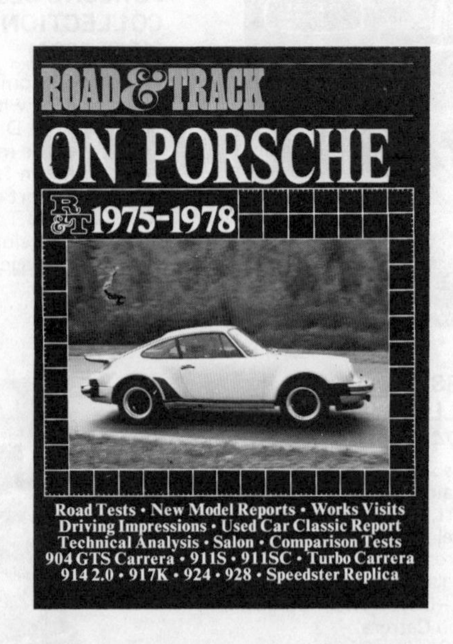

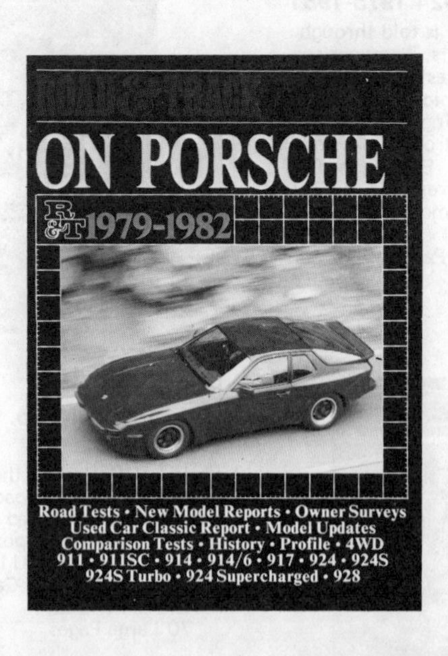

ROAD & TRACK ON PORSCHE 1975-1978

Some 23 major articles that appeared in this important US magazine between Sept. 1975 and Dec. 1978 are included in this book. Six road tests, three comparison tests, driving impressions, a technical analysis, a used car classic plus a 'Salon' article and a number of new model reports combine to form a valuable record for Porsche enthusiasts. Models covered include 904 GTS Carrera, 911S, 911SC, Turbo Carrera plus Group 4 and 5 Turbos, 914 2.0, 917K, 924, 928 and the Intermeccanica Speedster Replica.
88 Large Pages

ROAD & TRACK ON PORSCHE 1979-1982

Porsche progress during this period has been carefully monitored by Road & Track, through 9 road tests and 13 other stories. Informative comparisons are made between the 924, the Datsun 280 ZX, Corvette and Mazda RX-7GS and the 924 Turbo, Alfa Romeo GTV 6/2.5 and Datsun 280 ZX Turbo. Other articles cover a 911 Owner Survey, a 914 and 914/6 Used Car Classic, history, new model reports and a profile on Peter W Schutz. Models covered include 911, 911SC, 911 4WD, 914, 914/6, 917, 924, 924S, Butera's Supercharged 924, 924S Turbo and the 928.
88 Large Pages

In preparation Road & Track On Porsche 1972-1975

These soft-bound volumes in the 'Brooklands Books' Road & Track series consist of articles drawn from this important US journal. Fully illustrated with photographs and drawings, the articles contain road tests, driving impressions, performance figures and full specifications etc. Fascinating to read the books form an important reference source for owners, restorers, and other enthusiasts.

Some of the articles in the above books appear in the regular Brooklands Books reference series.

From specialist booksellers or, in case of difficulty, direct from the distributors:
BROOKLANDS BOOK DISTRIBUTION, 'HOLMERISE', SEVEN HILLS ROAD, COBHAM, SURREY KT11 1ES, ENGLAND. Telephone: Cobham (09326) 5051
MOTORBOOKS INTERNATIONAL, OSCEOLA, WISCONSIN 54020, USA.
Telephone: 715 294 3345 & 800 826 6600

BROOKLANDS BOOKS

PORSCHE BY YEARS

PORSCHE CARS 1960-1964

A collection of 31 fully illustrated road test reports and articles covering the following models: Super 75, Formula 2 car, 1600, Super 90, 360, Carrera 2, 365B, 1962 F1, 904, 901, 1600-S, 904 Carrera GTS and 356SC. Feature articles include 'Porsche Spring Playground', 'Porsche Prepares' (the F2 car), 'The Most Fantastic Grand Prix Car' and 'Class E Challenge' 100 large format pages.

PORSCHE CARS 1964-1968

This book consists of a total of 29 road test reports and articles and covers the 1600 SC, 904, 911, 365 SC, 912, Carrera 6, Super 75 Cabriolet, 911S, 912, 911L Sportomatic. Feature articles include: 'The Porsche and Me', 'Life with a Porsche', 'Driving Porsche's Pancake', 'Ferry Porsche II', 'Love Affair with a Bath Tub'. 100 large format pages. Fully illustrated.

PORSCHE CARS 1968-1972

A collection of road test reports and articles covering the following models: 911 Sportomatic, 911E, 914, 911T, 550 Spyder, 917. Feature articles include: 'The Toughest Porsche', Performance VW', 'Line up for '69', '1970 Plans', 'The Quick One', 'Valhalla is a VW-Porsche 914', 'Crayford VW-Porsche 914/4'. A total of 30 fully illustrated road tests and articles. 100 large pages.

PORSCHE CARS 1972-1975

Road tests, new model introductions. Articles on rallying, racing, Porsche history & covering the 911E, 911S, Targa, 911 speedster, 911T, 912E, 914 2.0, 928 V8, 904, Carrera RS and Turbo. 100 Large Pages.

PORSCHE CARS in the 60s

Road tests, analysis, new models, specifications also articles on engine swopping and Porsche competition cars. Models covered 365B 1600, 365B/Super 75, 365C, RS 60 Carrera GT, Super 90, 1600S, 901, 904 GT, 911, 911/2000, 911E, 911S, 911L, 911T, 911 Sportomatic, 912 Targa, 914, and 914/6. 100 Large Pages.

These soft-bound volumes in the 'Brooklands Books' series consist of reprints of original road test reports and other articles that appeared in leading motoring journals during the periods concerned. Fully illustrated with photographs and cut-away drawings, the articles contain road impressions, performance figures, specifications, etc. None of the articles appears in more than one book. Sources include Autocar, Autosport, Car, Car & Driver, Cars & Car Conversions, Motor, Motor Racing, Modern Motor, Road Test, Road & Track and Wheels. Fascinating to read, the books are also invaluable as sources of historical reference and as practical aids to enthusiasts who wish to restore their cars to original condition.

From specialist booksellers or, in case of difficulty, direct from the distributors: BROOKLANDS BOOK DISTRIBUTION, 'HOLMERISE', SEVEN HILLS ROAD, COBHAM, SURREY KT11 1ES, ENGLAND. Telephone: Cobham (09326) 5051 MOTORBOOKS INTERNATIONAL, OSCEOLA, WISCONSIN 54020, USA. Telephone: 715 294 3345 & 800 826 6600

LOTUS

LOTUS ELITE 1957-1964

The progress of the Lotus Elite is traced through 39 articles, of which 11 are road tests, others cover driving impressions, model introductions, touring, racing, a technical appraisal, a re-assessment, a track test, a used car test, a comparison vs. the Elan plus more recent historical articles. All models are dealt with including the twin-cam.
100 Large Pages.

LOTUS ELITE & ECLAT 1974-1981

The Lotus Elite & Eclat story is developed thro' 26 articles comparising of 9 Road Tests, 4 & 12000 m Reports, a comparison test vs the Porsche 924, driving impressions and information on buying a used car. Models covered include 501, 502, 503, 504, 521, 523, 2.2L, V8, M50, M52 and the Sprint.
100 Large Pages.

LOTUS ESPRIT 1975-1981

The views of Britain, the USA and Australia make up the 24 articles that trace the development of the Esprit from its introduction in 1975. Models covered include the S2, S3, B&C S2 Turbo, Esprit Turbo and the Essex. A total of 8 Road Tests, a comparison test vs the Lancia Monte Carlo, plus driving impressions, new model introductions and comprehensive specifications.
100 Large Pages.

LOTUS EUROPA 1966-1975

Road Tests (11) New model introductions, technical descriptions plus articles on tuning and touring make up the 31 stories in this book. All models are covered including the S2, S4, the twin cam, the Special plus the Racing version 47 and the GKN 47.
100 Large Pages.

LOTUS EUROPA COLLECTION No. 1

A further 26 articles bringing informed reports on the Europa from Australia, America, Britain and the Far East. Road Tests, driving impressions, specifications and new model introductions cover the period from 1966 to 1974 and include articles on the S1 and S2, the Else Europa, the Twin Cam and the Special as well as the 47 and the GKN V8 47.
70 Large Pages.

LOTUS SEVEN 1957-1980

A total of 28 articles tell the Lotus 7 Story from its introduction in 1957. They are made up of 11 Road Tests, an owners report, comparison tests plus stories on racing and history. Models include the America, 7, Super 7, Cosworth, Twin-Cam, IV, Seven Mazda, Caterham 7 & Twin-Cam.
100 Large Pages.

LOTUS SEVEN COLLECTION No. 1 (1957-1982)

The views of Britain, Australia, Malaysia and the U.S.A. are expressed in the 26 articles that trace the Lotus Seven story from its inception in 1958 to the Jubilee Model introduced to celebrate the model's 25th Anniversary. Articles cover 8 road tests, a touring trial, racing, history, assembly and new model introduction. Cars covered are the Seven, Super Seven, 1500, Series 4, 4SE, S4, plus the Caterham Super Sevens.
70 Large Pages.

LOTUS ELAN 1962-1973

The Elan 1500, 1600, Coupé, S2, 2+2, SE, S4, Sprint, +2S, 130S, 2S130, and S 130/5 are reported on. Some 8 Road Tests, a track test, a road research report and comparison test Vs the TR6 are included.
100 Large Pages.

LOTUS ELAN COLLECTION No. 1

This collection of articles covers the whole production life of the Elan from 1962 to 1974 and supplements the Elan marque book with completely different articles. Six Road Tests are included together with an owner survey a 24,000 mile report, and advice on buying a used Elan. Models covered include the 1500, 1600, S/E, Plus 2, Sprint, 2S130, and the S3.
70 Large Pages.

From specialist booksellers or, in case of difficulty, direct from the distributors:
BROOKLANDS BOOK DISTRIBUTION, 'HOLMERISE' SEVEN HILLS ROAD, COBHAM, SURREY KT11 1ES, ENGLAND. Telephone: Cobham (09326) 5051
MOTORBOOKS INTERNATIONAL, OSCEOLA, WISCONSIN 54020, USA.
Telephone: 715 294 3345 & 800 826 6600